First exam published in 2004.
Published by Leckie & Leckie Ltd, 3rd Floor, 4 Queen Street, Edinburgh EH2 1JE
tel: 0131 220 6831 fax: 0131 225 9987 enquiries@leckieandleckie.co.uk www.leckieandleckie.co.uk

ISBN 978-1-84372-686-9

A CIP Catalogue record for this book is available from the British Library.

Leckie & Leckie is a division of Huveaux plc.

Leckie & Leckie is grateful to the copyright holders, as credited at the back of the book, for permission to use their material.
Every effort has been made to trace the copyright holders and to obtain their permission for the use of copyright material.
Leckie & Leckie will gladly receive information enabling them to rectify any error or omission in subsequent editions.

Higher

Physics

2004 Exam

2005 Exam

2006 Exam

2007 Exam

2008 Exam

Leckie×Leckie

[BLANK]

X069/301

NATIONAL
QUALIFICATIONS
2004

FRIDAY, 28 MAY
1.00 PM – 3.30 PM

PHYSICS
HIGHER

Read Carefully

1 All questions should be attempted.

Section A (questions 1 to 20)

2 Check that the answer sheet is for Physics Higher (Section A).

3 Answer the questions numbered 1 to 20 on the answer sheet provided.

4 Fill in the details required on the answer sheet.

5 Rough working, if required, should be done only on this question paper, or on the first two pages of the answer book provided—**not** on the answer sheet.

6 For each of the questions 1 to 20 there is only **one** correct answer and each is worth 1 mark.

7 Instructions as to how to record your answers to questions 1–20 are given on page three.

Section B (questions 21 to 30)

8 Answer questions numbered 21 to 30 in the answer book provided.

9 Fill in the details on the front of the answer book.

10 Enter the question number clearly in the margin of the answer book beside each of your answers to questions 21 to 30.

11 Care should be taken to give an appropriate number of significant figures in the final answers to calculations.

SCOTTISH
QUALIFICATIONS
AUTHORITY

©

DATA SHEET
COMMON PHYSICAL QUANTITIES

Quantity	Symbol	Value	Quantity	Symbol	Value
Speed of light in vacuum	c	3.00×10^8 m s^{-1}	Mass of electron	m_e	9.11×10^{-31} kg
Magnitude of the charge on an electron	e	1.60×10^{-19} C	Mass of neutron	m_n	1.675×10^{-27} kg
Gravitational acceleration on Earth	g	9.8 m s^{-2}	Mass of proton	m_p	1.673×10^{-27} kg
Planck's constant	h	6.63×10^{-34} J s			

REFRACTIVE INDICES
The refractive indices refer to sodium light of wavelength 589 nm and to substances at a temperature of 273 K.

Substance	Refractive index	Substance	Refractive index
Diamond	2·42	Water	1·33
Crown glass	1·50	Air	1·00

SPECTRAL LINES

Element	Wavelength/nm	Colour	Element	Wavelength/nm	Colour
Hydrogen	656	Red	Cadmium	644	Red
	486	Blue-green		509	Green
	434	Blue-violet		480	Blue
	410	Violet			
	397	Ultraviolet			
	389	Ultraviolet			

Lasers		
Element	Wavelength/nm	Colour
Carbon dioxide	9550 } 10590 }	Infrared
Helium-neon	633	Red

Element	Wavelength/nm	Colour
Sodium	589	Yellow

PROPERTIES OF SELECTED MATERIALS

Substance	Density/ kg m^{-3}	Melting Point/ K	Boiling Point/ K
Aluminium	2.70×10^3	933	2623
Copper	8.96×10^3	1357	2853
Ice	9.20×10^2	273	. . . .
Sea Water	1.02×10^3	264	377
Water	1.00×10^3	273	373
Air	1·29	. . . .	. . . .
Hydrogen	9.0×10^{-2}	14	20

The gas densities refer to a temperature of 273 K and a pressure of 1.01×10^5 Pa.

SECTION A

For questions 1 to 20 in this section of the paper, an answer is recorded on the answer sheet by indicating the choice A, B, C, D or E by a stroke made in ink in the appropriate box of the answer sheet—see the example below.

EXAMPLE

The energy unit measured by the electricity meter in your home is the

    A ampere

    B kilowatt-hour

    C watt

    D coulomb

    E volt.

The correct answer to the question is B—kilowatt-hour. Record your answer by drawing a heavy vertical line joining the two dots in the appropriate box on your answer sheet in the column of boxes headed B. The entry on your answer sheet would now look like this:

If after you have recorded your answer you decide that you have made an error and wish to make a change, you should cancel the original answer and put a vertical stroke in the box you now consider to be correct. Thus, if you want to change an answer D to an answer B, your answer sheet would look like this:

If you want to change back to an answer which has already been scored out, you should enter a tick (✓) to the RIGHT of the box of your choice, thus:

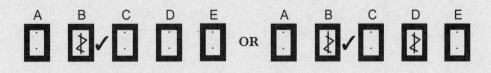

SECTION A

Answer questions 1–20 on the answer sheet.

1. A box is pulled along a level bench by a rope held at a constant angle of 40° to the horizontal as shown.

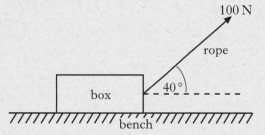

A constant force of 100 N is applied to the rope.

The box moves a distance of 10 m along the bench.

The work done on the box by the rope is

A   100 J

B   643 J

C   766 J

D   839 J

E   1000 J.

2. A stuntman on a motorcycle jumps a river which is 5·1 m wide. He lands on the edge of the far bank, which is 2·0 m lower than the bank from which he takes off.

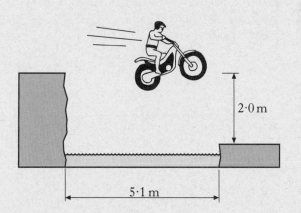

His minimum horizontal speed at take off is

A $2 \cdot 0 \, \mathrm{m \, s^{-1}}$

B $3 \cdot 2 \, \mathrm{m \, s^{-1}}$

C $5 \cdot 5 \, \mathrm{m \, s^{-1}}$

D $8 \cdot 0 \, \mathrm{m \, s^{-1}}$

E $9 \cdot 8 \, \mathrm{m \, s^{-1}}$.

3. A vehicle of mass 0·1 kg is moving to the right along a horizontal friction-free air track. A vehicle of mass 0·2 kg is moving to the left on the same track.

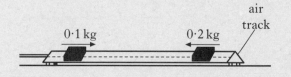

The vehicles collide and stick together.

Which of the following quantities is/are conserved in this collision?

  I   The total momentum

  II   The kinetic energy

  III   The total energy

A   I only

B   II only

C   I and II only

D   I and III only

E   II and III only

4. A cannon of mass 1200 kg fires a cannonball of mass 15 kg at a velocity of 60 m s⁻¹ East.

Assuming the force of friction is negligible, the velocity of the cannon just after firing is

A $0 \, \mathrm{m \, s^{-1}}$

B $0 \cdot 75 \, \mathrm{m \, s^{-1}}$ East

C $0 \cdot 75 \, \mathrm{m \, s^{-1}}$ West

D $6 \cdot 0 \, \mathrm{m \, s^{-1}}$ East

E $6 \cdot 0 \, \mathrm{m \, s^{-1}}$ West.

5. Car X is designed with a "crumple-zone" so that the front of the car collapses during impact as shown.

A similar car, Y, of equal mass is built without a crumple-zone. In a safety test both cars are driven at the same speed into identical walls.

Which of the following statements is/are true during the collisions?

I   The average force on car X is smaller than that on car Y.

II The time taken for car X to come to rest is greater than that for car Y.

III The change in momentum of car X is smaller than that of car Y.

A   I only

B   I and II only

C   I and III only

D   II and III only

E   I, II and III

6. A golf ball, initially at rest, is hit by a club. The graph of the force of the club on the ball against time is shown.

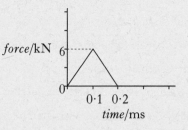

A different type of golf ball of the same size and mass is now hit with the same club. This ball moves off with the same velocity as the first ball.

Which graph shows the force of the club on the second ball against time?

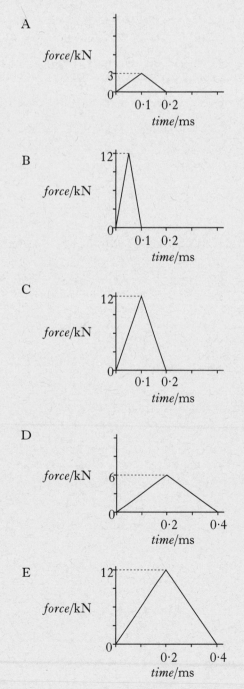

7. The density of steam at 100 °C is less than the density of water at 100 °C. The explanation for this is that when water changes to steam its particles

   A   move further apart

   B   move with greater speed

   C   have smaller mass

   D   are no longer joined together

   E   collide more often with each other.

8. In an experiment the following measurements and uncertainties are recorded.

   Temperature rise $= 10 \,°C \pm 1 \,°C$

   Heater current $= 5·0 \,A \pm 0·2 \,A$

   Heater voltage $= 12·0 \,V \pm 0·5 \,V$

   Time $= 100 \,s \pm 2 \,s$

   Mass of liquid $= 1·000 \,kg \pm 0·005 \,kg$

   The measurement which has the largest percentage uncertainty is the

   A   temperature rise

   B   heater current

   C   heater voltage

   D   time

   E   mass of liquid.

9. A balloon of volume of $6·0 \,m^3$ contains a fixed mass of gas at a temperature of 300 K and a pressure of 2·0 kPa. The gas is heated to 600 K and the pressure reduced to 1·0 kPa. The new volume of the gas is

   A $1·5 \,m^3$

   B $3·0 \,m^3$

   C $6·0 \,m^3$

   D $12·0 \,m^3$

   E $24·0 \,m^3$.

10. A student writes the following statements about electric fields.

    I   There is a force on a charge in an electric field.

    II When an electric field is applied to a conductor, the free electric charges in the conductor move.

    III Work is done when a charge is moved in an electric field.

    Which of the above statements is/are correct?

    A   I only

    B   II only

    C   I and II only

    D   I and III only

    E   I, II and III

11. A resistor and a capacitor are connected to identical a.c. supplies which provide constant voltage throughout their whole frequency range.

Which of the following pairs of graphs illustrates how the current varies with frequency in the two circuits shown?

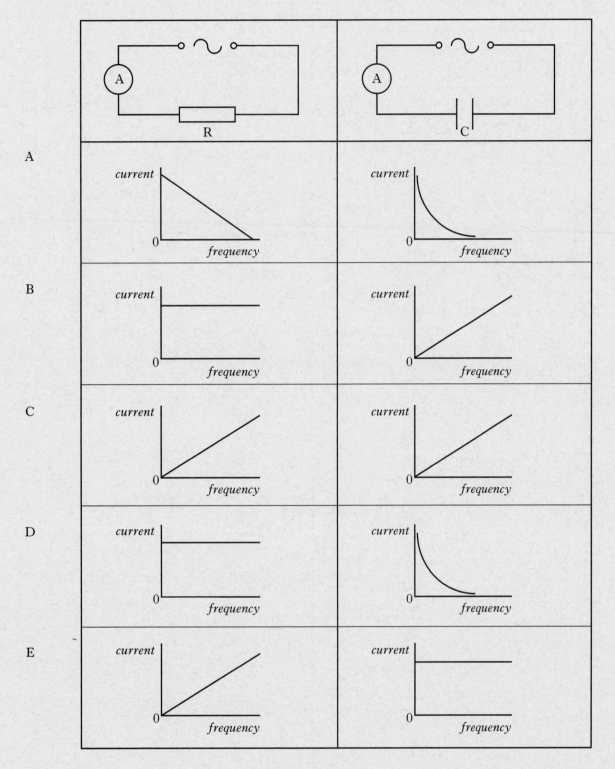

Page seven

[Turn over

12. The output from a signal generator is connected to the input terminals of an oscilloscope. The trace observed on the oscilloscope screen, the Y-gain setting and the time-base setting are shown in the diagram.

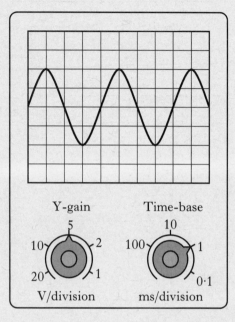

The frequency of the signal shown is calculated using the

A   Y-gain setting and the vertical height of the trace

B   Y-gain setting and the horizontal distance between the peaks of the trace

C   Y-gain setting and time-base setting

D   time-base setting and the vertical height of the trace

E   time-base setting and the horizontal distance between the peaks of the trace.

13. Which of the following statements is/are true for an ideal op-amp?

   I   It has infinite input resistance.

   II   Both inputs are at the same potential.

   III   The input current to the op-amp is zero.

   A   I only

   B   II only

   C   I and II only

   D   II and III only

   E   I, II and III

14. Two identical loudspeakers, L_1 and L_2, are operated at the same frequency and in phase with each other. An interference pattern is produced.

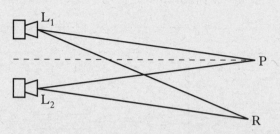

At position P, which is the same distance from both loudspeakers, there is a maximum intensity.

The next maximum intensity is at position R, where $L_1R = 5\cdot6\,m$ and $L_2R = 5\cdot3\,m$.

The speed of sound is $340\,m\,s^{-1}$.

The frequency of the sound emitted by the loudspeakers is given by

A $\dfrac{5\cdot6-5\cdot3}{340}$ Hz

B $\dfrac{340}{5\cdot6+5\cdot3}$ Hz

C $\dfrac{340}{5\cdot6-5\cdot3}$ Hz

D $340\times(5\cdot6-5\cdot3)$ Hz

E $340\times(5\cdot6+5\cdot3)$ Hz.

15. Ultraviolet radiation is incident on a clean zinc plate. Photoelectrons are ejected.

The clean zinc plate is replaced by a different metal which has a lower work function. The same intensity of ultraviolet radiation is incident on this metal.

Compared to the zinc plate, which of the following statements is/are true for the new metal?

   I   The maximum speed of the photoelectrons is greater.

   II   The maximum kinetic energy of the photoelectrons is greater.

   III   There are more photoelectrons ejected per second.

   A   I only

   B   II only

   C   III only

   D   I and II only

   E   I, II and III

16. An atom has the energy levels shown.

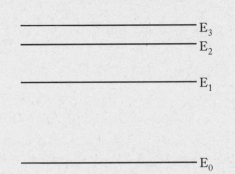

$$E_3$$
$$E_2$$

$$E_1$$

$$E_0$$

Electron transitions occur between all of these levels to produce emission lines in the spectrum of this atom.

How many emission lines are produced by transitions between these energy levels?

A   3

B   4

C   5

D   6

E   7

17. Materials are "doped" to produce n-type semiconductor material.

In n-type semiconductor material

A   the majority charge carriers are electrons

B   the majority charge carriers are neutrons

C   the majority charge carriers are protons

D   there are more protons than neutrons

E   there are more electrons than neutrons.

18. A student writes the following statements about the decay of radionuclides.

I   During alpha emission a particle consisting of 2 protons and 4 neutrons is emitted from a nucleus.

II   During beta emission a fast moving electron is emitted from a nucleus.

III   During gamma emission a high energy photon is emitted from a nucleus.

Which of these statements is/are true?

A   II only

B   I and II only

C   I and III only

D   II and III only

E   I, II and III

19. A radiation technician works 150 hours each month in an area exposed to radiation from a neutron beam. The quality factor for this radiation is 3. The technician receives an absorbed dose rate of $10\,\mu Gy\,h^{-1}$ from this radiation.

In a period of 5 months the total dose equivalent received by the technician is

A $2 \cdot 50 \times 10^{-2}\,Sv$

B $2 \cdot 25 \times 10^{-2}\,Sv$

C $1 \cdot 50 \times 10^{-2}\,Sv$

D $1 \cdot 00 \times 10^{-2}\,Sv$

E $0 \cdot 75 \times 10^{-2}\,Sv.$

20. A Geiger counter records a corrected count-rate of 1000 counts per second when it is placed a distance of 400 mm from a radioactive source.

A sheet of metal is placed between the source and the counter. The half value thickness of the metal for radiation from the source is 20 mm.

The corrected count-rate is now 125 counts per second.

The thickness of the metal sheet is

A   25 mm

B   40 mm

C   60 mm

D   80 mm

E   160 mm.

[Turn over

SECTION B

Write your answers to questions 21 to 30 in the answer book.

Marks

21. (a) State the difference between speed and velocity.

    1

    (b) During a tall ships race, a ship called the Mir passes a marker buoy X and sails due West (270). It sails on this course for 30 minutes at a speed of $10.0\,km\,h^{-1}$, then changes course to 20° West of North (340). The Mir continues on this new course for $1\frac{1}{2}$ hours at a speed of $8.0\,km\,h^{-1}$ until it passes marker buoy Y.

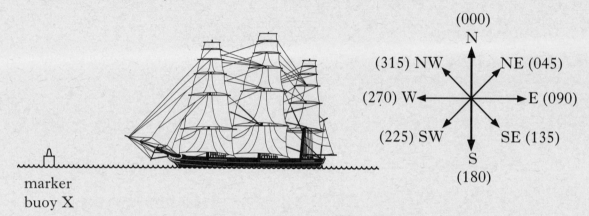

marker
buoy X

    (i) Show that the Mir travels a total distance of 17 km between marker buoys X and Y.

    (ii) By scale drawing or otherwise, find the displacement from marker buoy X to marker buoy Y.

    (iii) Calculate the average velocity, in $km\,h^{-1}$, of the Mir between marker buoys X and Y.

    6

    (c) A second ship, the Leeuvin, passes marker buoy X 15 minutes after the Mir and sails directly for marker buoy Y at a speed of $7.5\,km\,h^{-1}$.

    Show by calculation which ship first passes marker buoy Y.

    2

    (9)

Marks

22. A train of mass 7.5×10^5 kg is travelling at $60 \, \text{m s}^{-1}$ along a straight horizontal track.

The brakes are applied and the train decelerates uniformly to rest in a time of 40 s.

(a) (i) Calculate the distance the train travels between the brakes being applied and the train coming to rest.

   (ii) Calculate the force required to bring the train to rest in this time. **4**

(b) Part of the train's braking system consists of an electrical circuit as shown in the diagram.

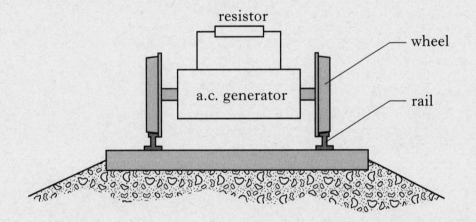

While the train is braking, the wheels drive an a.c. generator which changes kinetic energy into electrical energy. This electrical energy is changed into heat in a resistor. The r.m.s. current in the resistor is 2.5×10^3 A and the resistor produces 8.5 MJ of heat each second.

Calculate the peak voltage across the resistor. **3**

(7)

[Turn over

Marks

23. A crane barge is used to place part of an oil well, called a manifold, on the seabed.

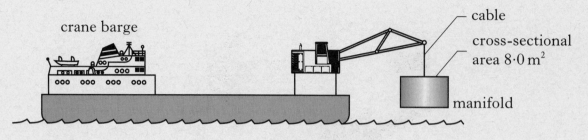

crane barge

cable

cross-sectional area $8.0\,\text{m}^2$

manifold

The manifold is a cylinder of uniform cross-sectional area $8.0\,\text{m}^2$ and mass $5.0 \times 10^4\,\text{kg}$. The mass of the cable may be ignored.

(a) Calculate the tension in the cable when the manifold is held stationary above the surface of the water.

1

(b) The manifold is lowered into the water and then held stationary just below the surface as shown.

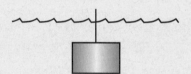

   (i) Draw a sketch showing all the forces acting vertically on the manifold. Name each of these forces.

   (ii) The tension in the cable is now $2.5 \times 10^5\,\text{N}$.

      Show that the difference in pressure between the top and bottom surfaces of the manifold is $3.0 \times 10^4\,\text{Pa}$.

4

(c) The manifold is now lowered to a greater depth.

What effect does this have on the difference in pressure between the top and bottom surfaces of the manifold?

You must justify your answer.

2

(7)

Marks

24. A student sets up the circuit shown.

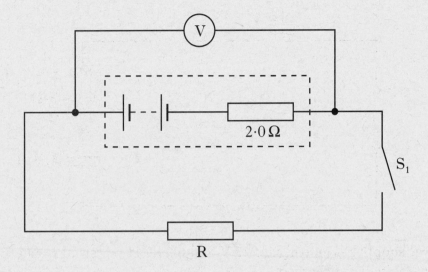

The internal resistance of the battery is $2 \cdot 0\,\Omega$.

With S_1 open, the student notes that the reading on the voltmeter is $9 \cdot 0\,V$.

The student closes S_1 and notes that the reading on the voltmeter is now $7 \cdot 8\,V$.

(a) (i) Calculate the resistance of resistor R.

    (ii) Explain why the reading on the voltmeter decreases when S_1 is closed.   **4**

(b) The student adds a $30\,\Omega$ resistor and a switch S_2 to the circuit as shown.

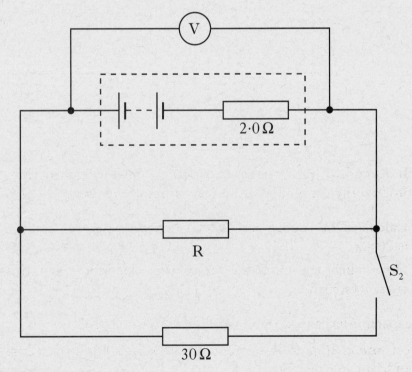

The student now closes S_2.

Explain what happens to the reading on the voltmeter.   **2**

                                              (6)

Marks

25. In an experiment, the circuit shown is used to investigate the charging of a capacitor.

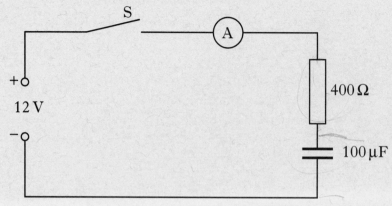

The power supply has an e.m.f. of 12 V and negligible internal resistance. The capacitor is initially uncharged.

Switch S is closed and the current measured during charging. The graph of charging current against time is shown in figure 1.

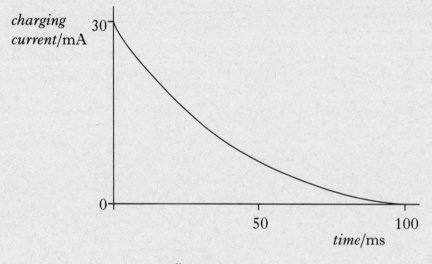

figure 1

(a) Sketch a graph of the voltage across the capacitor against time until the capacitor is fully charged. Numerical values are required on both axes.　　2

(b)　(i)　Calculate the voltage across the capacitor when the charging current is 20 mA.

　　(ii)　How much energy is stored in the capacitor when the charging current is 20 mA?　　4

(c)　The capacitor has a maximum working voltage of 12 V.

　　Suggest **one** change to this circuit which would allow an initial charging current of greater than 30 mA.　　1

Marks

25. **(continued)**

(*d*) The 100 μF capacitor is now replaced by an uncharged capacitor of unknown capacitance and the experiment is repeated. The graph of charging current against time for this capacitor is shown in figure 2.

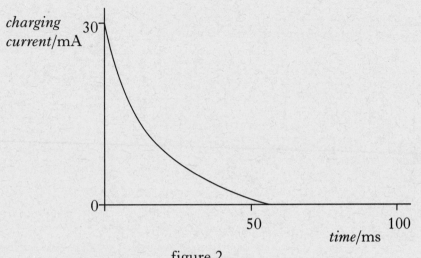

figure 2

By comparing figure 2 with figure 1, determine whether the capacitance of this capacitor is greater than, equal to or less than 100 μF.

You must justify your answer.

2

(9)

[Turn over

Marks

26. The circuit shown is designed for an alarm system.

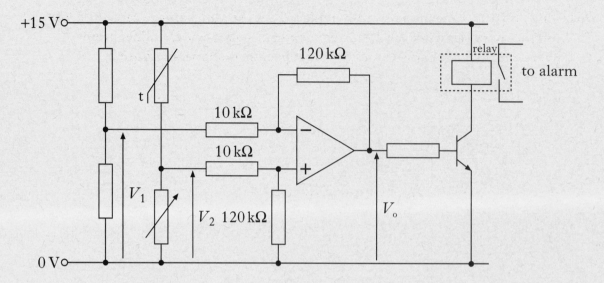

Voltage V_1 is 7·50 V.

When the temperature increases, the resistance of the thermistor decreases.

(a) At a temperature of 35 °C, voltage V_2 is 7·52 V.

Calculate the output voltage V_o at this temperature. 2

(b) When the temperature rises, V_o increases and the alarm switches on.

Explain how the circuit operates to switch on the alarm. 2

(c) The alarm is on when V_o is greater than or equal to 0·72 V.

The graph of the temperature against voltage V_2 is shown.

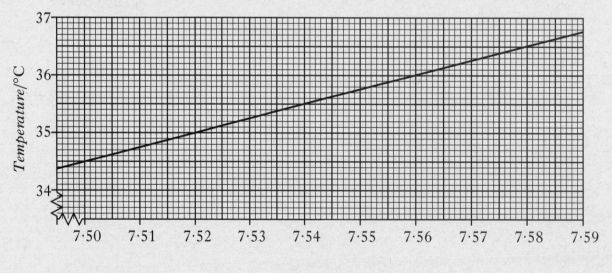

Using information from the graph, determine the minimum temperature at which the alarm switches on. 2

(6)

Marks

27. A decorative lamp has a transparent liquid in the space above a bulb. Light from the bulb passes through rotating coloured filters giving red or blue light in the liquid.

(a) A ray of red light is incident on the liquid surface as shown.

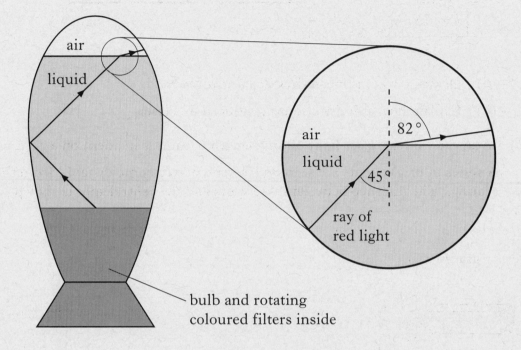

bulb and rotating
coloured filters inside

(i) Calculate the refractive index of the liquid for the red light.

(ii) A ray of blue light is incident on the liquid surface at the same angle as the ray of red light.

The refractive index of the liquid for blue light is greater than that for red light. Is the angle of refraction greater than, equal to or less than 82° for the blue light?

You must explain your answer. **3**

(b) A similar lamp contains a liquid which has a refractive index of 1·44 for red light. A ray of red light in the liquid is incident on the surface at an angle of 45° as before.

Sketch a diagram to show the path of this ray after it is incident on the liquid surface.

Mark on your diagram the values of all appropriate angles.

All relevant calculations must be shown. **3**

(6)

[Turn over

Marks

28. The term LASER is short for "Light Amplification by the Stimulated Emission of Radiation".

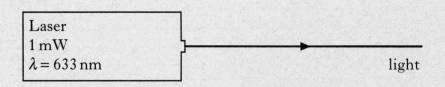

(*a*) (i) Describe what is meant by *Stimulated Emission*.

    (ii) Explain how amplification is produced in a laser.   **3**

(*b*) In an experiment, laser light of wavelength 633 nm is incident on a grating.

A series of bright spots are seen on a screen placed some distance from the grating. The distance between these spots and the central spot is shown.

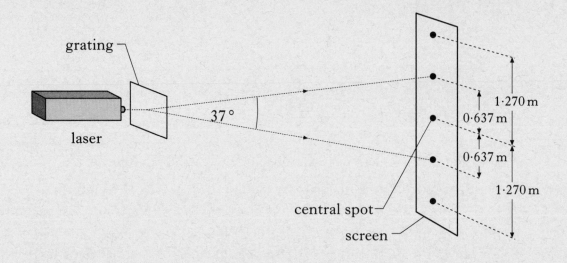

Calculate the number of lines per metre on the grating.   **3**

(*c*) The laser is replaced with another laser and the experiment repeated. With this laser the bright spots are closer together.

How does the wavelength of the light from this laser compare with that from the original laser?

You must justify your answer.   **2**

    (8)

Marks

29. An LED consists of a p-n junction as shown.

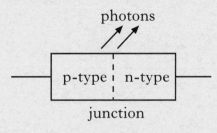

(*a*) Copy the diagram and add a battery so that the p-n junction is forward-biased.

1

(*b*) Using the terms *electrons, holes* and *photons*, explain how light is produced at the p-n junction of the LED.

1

(*c*) The LED emits photons, of energy $3\cdot68 \times 10^{-19}$ J.

   (i) Calculate the wavelength of a photon of light from this LED.

   (ii) Calculate the minimum potential difference across the p-n junction when it emits photons.

4

(6)

[Turn over for Question 30 on *Page twenty*

Marks

30. A ship is powered by a nuclear reactor.

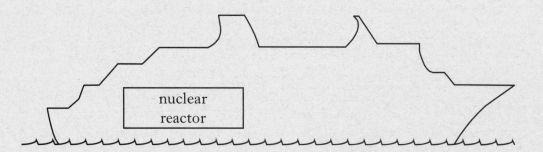

One reaction that takes place in the core of the nuclear reactor is represented by the statement below.

$$^{235}_{92}U + ^{1}_{0}n \rightarrow ^{140}_{58}Ce + ^{94}_{40}Zr + 2^{1}_{0}n + 6^{0}_{-1}e$$

(a) The symbol for the Uranium nucleus is $^{235}_{92}U$.

What information about the nucleus is provided by the following numbers?

(i) 92

(ii) 235   2

(b) Describe how neutrons produced during the reaction can cause further nuclear reactions.   1

(c) The masses of particles involved in the reaction are shown in the table.

Particles	Mass/kg
$^{235}_{92}U$	$390 \cdot 173 \times 10^{-27}$
$^{140}_{58}Ce$	$232 \cdot 242 \times 10^{-27}$
$^{94}_{40}Zr$	$155 \cdot 884 \times 10^{-27}$
$^{1}_{0}n$	$1 \cdot 675 \times 10^{-27}$
$^{0}_{-1}e$	negligible

Calculate the energy released in the reaction.   3

                                                                                     (6)

[END OF QUESTION PAPER]

[BLANK PAGE]

X069/301

NATIONAL
QUALIFICATIONS
2005

TUESDAY, 24 MAY
1.00 PM – 3.30 PM

PHYSICS
HIGHER

Read Carefully

1   All questions should be attempted.

Section A (questions 1 to 20)

2   Check that the answer sheet is for Physics Higher (Section A).

3   Check that the answer sheet you have been given has **your name**, **date of birth**, **SCN** (Scottish Candidate Number) and **Centre Name** printed on it.

 Do not change any of these details.

4   If any of this information is wrong, tell the Invigilator immediately.

5   If this information is correct, **print** your name and seat number in the boxes provided.

6   Use **black** or **blue ink** for your answers. **Do not use red ink**.

7   There is **only one correct** answer to each question.

8   Any rough working should be done on the question paper or the rough working sheet, **not** on your answer sheet.

9   At the end of the exam, put the **answer sheet for Section A inside the front cover of your answer book**.

10 Instructions as to how to record your answers to questions 1–20 are given on page three.

Section B (questions 21 to 30)

11 Answer questions numbered 21 to 30 in the answer book provided.

12 Fill in the details on the front of the answer book.

13 Enter the question number clearly in the margin of the answer book beside each of your answers to questions 21 to 30.

14 Care should be taken to give an appropriate number of significant figures in the final answers to calculations.

SCOTTISH
QUALIFICATIONS
AUTHORITY

DATA SHEET
COMMON PHYSICAL QUANTITIES

Quantity	Symbol	Value	Quantity	Symbol	Value
Speed of light in vacuum	c	3.00×10^8 m s^{-1}	Mass of electron	m_e	9.11×10^{-31} kg
Magnitude of the charge on an electron	e	1.60×10^{-19} C	Mass of neutron	m_n	1.675×10^{-27} kg
Gravitational acceleration on Earth	g	9.8 m s^{-2}	Mass of proton	m_p	1.673×10^{-27} kg
Planck's constant	h	6.63×10^{-34} J s			

REFRACTIVE INDICES
The refractive indices refer to sodium light of wavelength 589 nm and to substances at a temperature of 273 K.

Substance	Refractive index	Substance	Refractive index
Diamond	2·42	Water	1·33
Crown glass	1·50	Air	1·00

SPECTRAL LINES

Element	Wavelength/nm	Colour	Element	Wavelength/nm	Colour
Hydrogen	656	Red	Cadmium	644	Red
	486	Blue-green		509	Green
	434	Blue-violet		480	Blue
	410	Violet			
	397	Ultraviolet			
	389	Ultraviolet			
Sodium	589	Yellow			

Lasers		
Element	Wavelength/nm	Colour
Carbon dioxide	9550 } 10590 }	Infrared
Helium-neon	633	Red

PROPERTIES OF SELECTED MATERIALS

Substance	Density/ kg m^{-3}	Melting Point/ K	Boiling Point/ K
Aluminium	2.70×10^3	933	2623
Copper	8.96×10^3	1357	2853
Ice	9.20×10^2	273	....
Sea Water	1.02×10^3	264	377
Water	1.00×10^3	273	373
Air	1·29	....	....
Hydrogen	9.0×10^{-2}	14	20

The gas densities refer to a temperature of 273 K and a pressure of 1.01×10^5 Pa.

SECTION A

For questions 1 to 20 in this section of the paper the answer to each question is either A, B, C, D or E. Decide what your answer is, then put a horizontal line in the space provided—see the example below.

EXAMPLE

The energy unit measured by the electricity meter in your home is the

    A ampere

    B kilowatt-hour

    C watt

    D coulomb

    E volt.

The correct answer is **B**—kilowatt-hour. The answer **B** has been clearly marked with a horizontal line (see below).

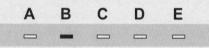

Changing an answer

If you decide to change your answer, cancel your first answer by putting a cross through it (see below) and fill in the answer you want. The answer below has been changed to **B**.

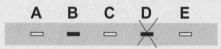

If you then decide to change back to an answer you have already scored out, put a tick (✓) to the **right** of the answer you want, as shown below:

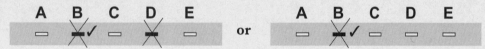

[Turn over

SECTION A

Answer questions 1–20 on the answer sheet.

1. A car travels from X to Y and then from Y to Z as shown.

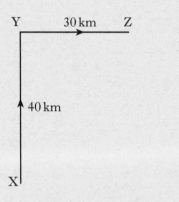

It takes one hour to travel from X to Y. It also takes one hour to travel from Y to Z.

Which row in the following table shows the magnitudes of the displacement, average speed and average velocity for the complete journey?

	Displacement (km)	Average speed (km h^{-1})	Average velocity (km h^{-1})
A	50	35	25
B	70	35	25
C	50	35	35
D	70	70	50
E	50	70	25

2. An object has a constant acceleration of $3 \, \text{m s}^{-2}$. This means that the

A   distance travelled by the object increases by 3 metres every second

B   displacement of the object increases by 3 metres every second

C   speed of the object is $3 \, \text{m s}^{-1}$ every second

D   velocity of the object is $3 \, \text{m s}^{-1}$ every second

E   velocity of the object increases by $3 \, \text{m s}^{-1}$ every second.

3. A car of mass 1200 kg pulls a horsebox of mass 700 kg along a straight, horizontal road. They have an acceleration of $2 \cdot 0 \, \text{m s}^{-2}$.

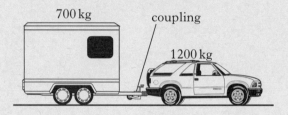

Assuming that the frictional forces are negligible, the tension in the coupling between the car and the horsebox is

A   500 N

B   700 N

C   1400 N

D   2400 N

E   3800 N.

4. A mass of 2 kg slides along a frictionless surface at $10 \, \text{m s}^{-1}$ and collides with a stationary mass of 10 kg.

before impact

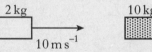

After the collision, the 2 kg mass rebounds at $5 \, \text{m s}^{-1}$ and the 10 kg mass moves off at $3 \, \text{m s}^{-1}$.

after impact

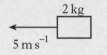

Which row in the following table is correct?

	Momentum of system	Kinetic energy of system	Type of collision
A	conserved	conserved	elastic
B	conserved	not conserved	inelastic
C	conserved	not conserved	elastic
D	not conserved	not conserved	inelastic
E	not conserved	not conserved	elastic

5. A golfer hits a ball of mass $5 \cdot 0 \times 10^{-2}$ kg with a golf club. The ball leaves the tee with a velocity of $80 \, \text{m s}^{-1}$. The club is in contact with the ball for a time of $0 \cdot 10$ s.

   The average force exerted by the club on the ball is

   A $6 \cdot 25 \times 10^{-4}$ N

   B $0 \cdot 025$ N

   C $0 \cdot 4$ N

   D   4 N

   E   40 N.

6. A solid at a temperature of $-20 \, ^\circ\text{C}$ is heated until it becomes a liquid at $70 \, ^\circ\text{C}$.

   The temperature change in kelvin is

   A   50 K

   B   90 K

   C   343 K

   D   363 K

   E   596 K.

7. One volt is

   A   one coulomb per joule

   B   one joule coulomb

   C   one joule per coulomb

   D   one joule per second

   E   one coulomb per second.

8. A potential difference of 5000 V is applied between two metal plates. The plates are $0 \cdot 10$ m apart. A charge of $+2 \cdot 0$ mC is released from rest at the positively charged plate as shown.

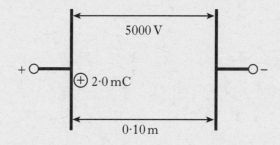

   The kinetic energy of the charge just before it hits the negative plate is

   A $4 \cdot 0 \times 10^{-7}$ J

   B $2 \cdot 0 \times 10^{-4}$ J

   C $5 \cdot 0$ J

   D   10 J

   E   500 J.

9. An a.c. signal is displayed on an oscilloscope screen. The Y-gain and time-base controls are set as shown.

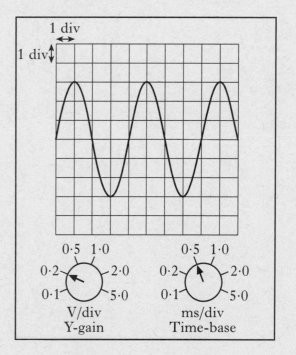

   The frequency of the signal is

   A $0 \cdot 50$ Hz

   B $1 \cdot 25$ Hz

   C $2 \cdot 00$ Hz

   D   200 Hz

   E   500 Hz.

10. A capacitor is connected to a circuit as shown.

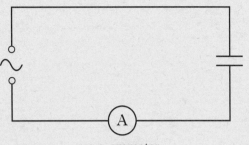

a.c. ammeter

The alternating supply has a constant peak voltage but its frequency can be varied.

The frequency is steadily increased from 50 Hz to 5000 Hz. The reading on the a.c. ammeter

A   remains constant

B   decreases steadily

C   increases steadily

D   increases then decreases

E   decreases then increases.

11. An amplifier circuit is shown.

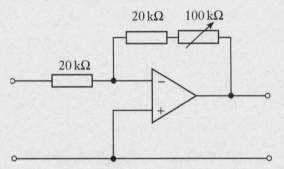

The variable resistor can be adjusted from zero to 100 kΩ. This allows the voltage gain to be altered over the range

A   zero to one

B   zero to five

C   zero to six

D   one to five

E   one to six.

12. A student sets up the following circuit.

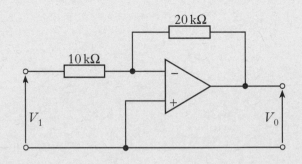
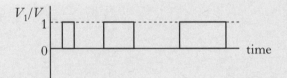

The graph below shows how the input voltage V_1 varies with time.

Which of the following graphs shows how the output voltage V_0 varies with time?

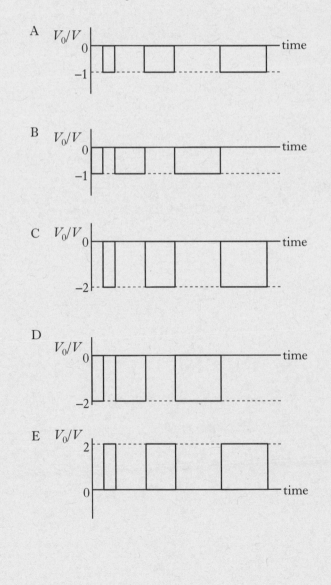

13. The apparatus used to investigate the relationship between light intensity I and distance d from a point source is shown.

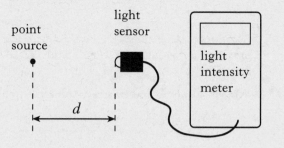

The experiment is carried out in a darkened room.

Which of the following expressions gives a constant value?

A $I \times d$

B $I \times d^2$

C $\dfrac{I}{d}$

D $\dfrac{I}{d^2}$

E $I \times \sqrt{d}$

14. Microwaves of frequency $2 \cdot 0 \times 10^{10}$ Hz travel through air with a speed of $3 \cdot 0 \times 10^8 \, \text{m s}^{-1}$. On entering a bath of oil, the speed reduces to $1 \cdot 5 \times 10^8 \, \text{m s}^{-1}$.

The frequency of the microwaves in the oil is

A $1 \cdot 0 \times 10^{10}$ Hz

B $2 \cdot 0 \times 10^{10}$ Hz

C $4 \cdot 0 \times 10^{10}$ Hz

D $3 \cdot 0 \times 10^{18}$ Hz

E $6 \cdot 0 \times 10^{18}$ Hz.

15. Which graph shows the relationship between frequency f and wavelength λ of photons of electromagnetic radiation?

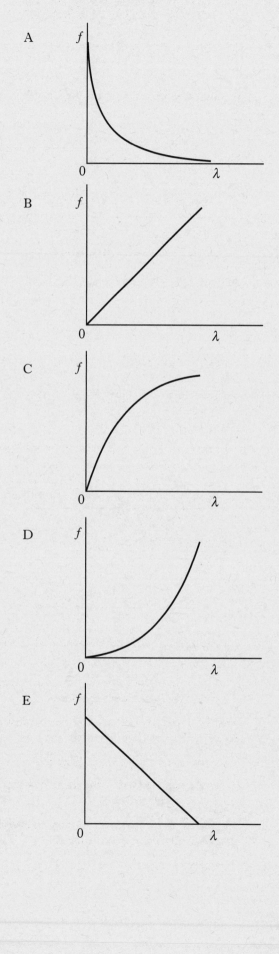

16. A liquid and a solid have the same refractive index.

    What happens to the speed and the wavelength of light passing from the liquid into the solid?

	Speed	Wavelength
A	stays the same	stays the same
B	decreases	decreases
C	decreases	increases
D	increases	increases
E	increases	decreases

17. The intensity of light can be measured in

    A W

    B $W\,m^{-1}$

    C $W\,m$

    D $W\,m^{-2}$

    E $W\,m^{2}$.

18. The diagram below represents part of the process of stimulated emission in a laser.

    Which of the following statements best describes the emitted radiation?

    A Out of phase and emitted in the same direction as the incident radiation

    B Out of phase and emitted in the opposite direction to the incident radiation

    C In phase and emitted in all directions.

    D In phase and emitted in the same direction as the incident radiation

    E In phase and emitted in the opposite direction to the incident radiation

19. Part of a radioactive decay series is shown.

    $$^{P}_{Q}Bi \xrightarrow[\text{decay}]{\beta} \, ^{R}_{S}Po \xrightarrow[\text{decay}]{\alpha} \, ^{208}_{82}Pb$$

    A bismuth nucleus emits a beta particle and its product, a polonium nucleus, emits an alpha particle.

    Which numbers are represented by P, Q, R and S?

	P	Q	R	S
A	212	85	212	84
B	212	83	212	84
C	212	85	208	83
D	210	83	208	81
E	210	85	210	84

20. The equation below represents a nuclear reaction.

    $$^{235}_{92}U + ^{1}_{0}n \longrightarrow \, ^{92}_{36}Kr + ^{141}_{56}Ba + ^{1}_{0}n + ^{1}_{0}n + ^{1}_{0}n$$

    It is an example of

    A nuclear fusion

    B alpha particle emission

    C beta particle emission

    D induced nuclear fission

    E spontaneous nuclear fission.

[SECTION B begins on *Page ten*]

[Turn over

Marks

SECTION B

Write your answers to questions 21 to 30 in the answer book.

21. (*a*) A student uses the apparatus shown to measure the average acceleration of a trolley travelling down a track.

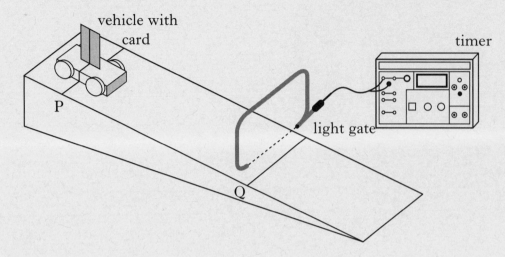

The line on the trolley is aligned with line P on the track.

The trolley is released from rest and allowed to run down the track.

The timer measures the time for the card to pass through the light gate.

This procedure is repeated a number of times and the results shown below.

0·015 s   0·013 s   0·014 s   0·019 s   0·017 s   0·018 s

(i) Calculate:

(A) the mean time for the card to pass through the light gate;   **1**

(B) the approximate absolute random uncertainty in this value.   **1**

(ii) The length of the card is 0·020 m and the distance PQ is 0·60 m.

Calculate the acceleration of the trolley (an uncertainty in this value is not required).   **3**

Marks

21. (continued)

(*b*) The light gate consists of a lamp shining onto a photodiode.

The photodiode forms part of the circuit shown.

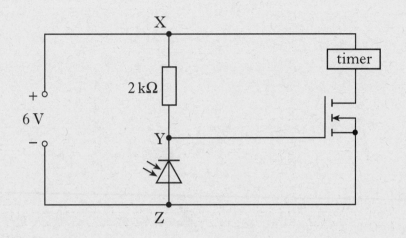

    (i)   In which mode is the photodiode operating? **1**

    (ii) Explain why the timer only operates while the light beam is broken. **2**

(8)

[Turn over

Marks

22. A "giant catapult" is part of a fairground ride.

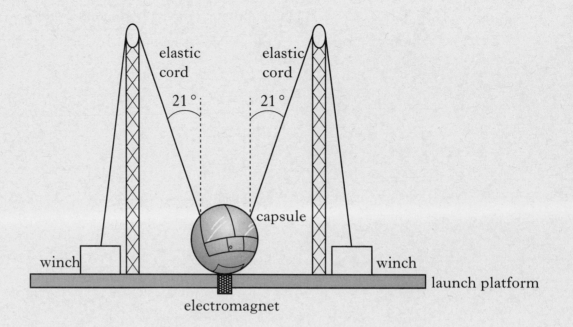

Two people are strapped into a capsule. The capsule and the occupants have a combined mass of 236 kg.

The capsule is held stationary by an electromagnet while the tension in the elastic cords is increased using the winches.

The mass of the elastic cords and the effects of air resistance can be ignored.

(a) When the tension in each cord reaches $4 \cdot 5 \times 10^3$ N the electromagnet is switched off and the capsule and occupants are propelled vertically upwards.

   (i)   Calculate the vertical component of the force exerted by **each** cord just before the capsule is released.   **1**

   (ii)   Calculate the initial acceleration of the capsule.   **3**

   (iii)   Explain why the acceleration of the capsule decreases as it rises.   **1**

(b) Throughout the ride the occupants remain upright in the capsule.

A short time after release the occupants feel no force between themselves and the seats.

Explain why this happens.   **1**

    (6)

23. A polystyrene float is held with its base 2·0 m below the surface of a swimming pool.

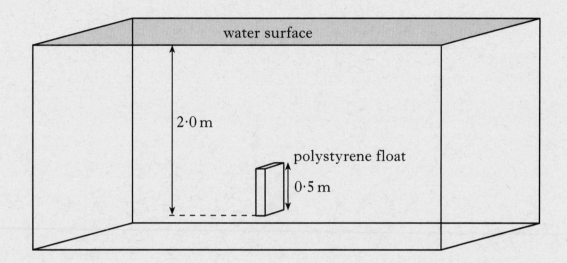

(a) The float has a mass of 12 g and its dimensions are 0·50 m × 0·30 m × 0·10 m.

Calculate the density of the float.   **2**

(b) Explain why a buoyancy force acts on the float.   **2**

(c) The float is released and accelerates towards the surface. Taking into account the resistance of the water, state what happens to the acceleration of the float as it approaches the surface. You must justify your answer.   **2**

(d) Another float made of a more dense material with the same dimensions is now held at the same position in the pool.

The float is released as in part (c).

State how the initial acceleration of this float compares with the polystyrene float. You must justify your answer.   **1**

(7)

[Turn over

Marks

24. The apparatus used to investigate the relationship between volume and temperature of a fixed mass of air is shown.

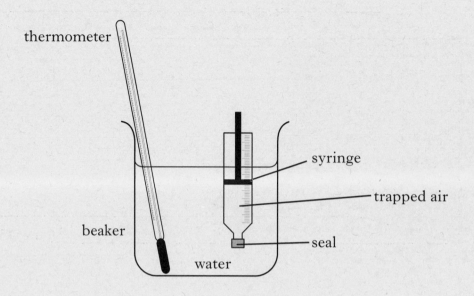

The volume of the trapped air is read from the scale on the syringe.

The temperature of the trapped air is altered by heating the water in the beaker. It is assumed that the temperature of the air in the syringe is the same as that of the surrounding water. The pressure of the trapped air is constant during the investigation.

(a) Readings of volume and temperature for the trapped air are shown.

Temperature/°C	25	50	75	100
Volume/ml	20·6	22·6	24·0	25·4

   (i) Using **all** the data, establish the relationship between temperature and volume for the trapped air.

2

   (ii) Calculate the volume of the trapped air when the temperature of the water is 65 °C.

2

   (iii) Use the kinetic model of gases to explain the change in volume as the temperature increases in this investigation.

2

Marks

24. (continued)

(b) An alternative to measuring the volume using the scale on the syringe, is to connect the piston of the syringe to a variable resistor.

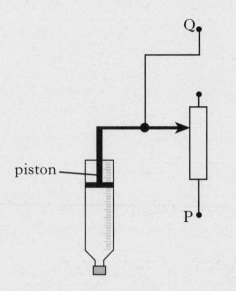

The variable resistor forms part of the circuit shown.

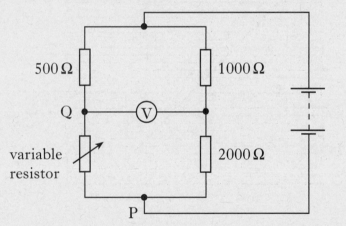

The reading on the voltmeter is 0 V when the temperature of the air in the syringe is 50 °C.

   (i) Calculate the resistance of the variable resistor at this temperature.   **2**

   (ii) The temperature of the gas in the syringe changes from just below to just above 50 °C. This causes the resistance of the variable resistor to change by a small amount.

       Sketch a graph of the reading on the centre-zero voltmeter against the change in resistance of the variable resistor. Numerical values are not required on either axis.   **1**

                                                       (9)

[Turn over

Marks

25. A student sets up the following circuit to find the e.m.f. *E* and the internal resistance *r* of a battery.

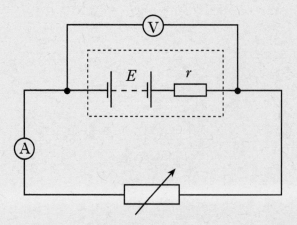

Readings from the voltmeter and ammeter are used to plot the following graph.

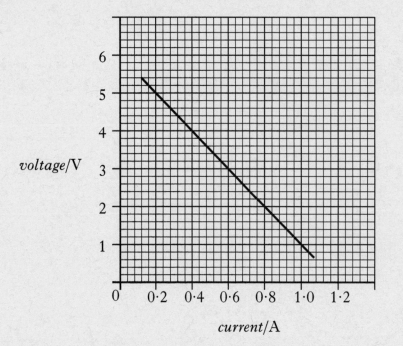

(a) What is meant by the term *e.m.f.*?   1

(b) (i) Use the graph to determine:

    (A) the e.m.f.;   1

    (B) the internal resistance of the battery.   2

(ii) Show that the variable resistor has a value of $15\,\Omega$ when the current is $0.30\,A$.   1

Marks

25. **(continued)**

(c) Without adjusting the variable resistor, a $30\,\Omega$ resistor is connected in parallel with it.

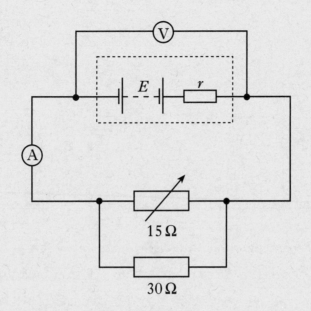

Calculate the new reading on the ammeter.

2

(7)

[Turn over

Marks

26. A student investigates the charging and discharging of a 2200 µF capacitor using the circuit shown.

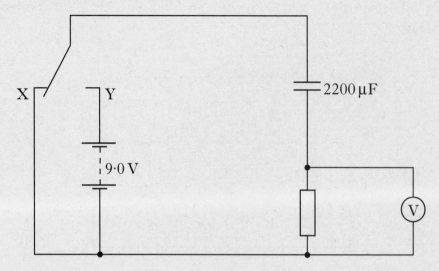

The 9·0 V battery has negligible internal resistance.

Initially the capacitor is uncharged and the switch is at position X.

The switch is then moved to position Y and the capacitor charges fully in 1·5 s.

(a) (i) Sketch a graph of the p.d. across the **resistor** against time while the capacitor charges. Appropriate numerical values are required on both axes.   2

  (ii) The resistor is replaced with one of higher resistance.

      Explain how this affects the time taken to fully charge the capacitor.   1

 (iii) At one instant during the charging of the capacitor the reading on the voltmeter is 4·0 V.

      Calculate the charge stored by the capacitor at this instant.   3

(b) Using the same circuit in a later investigation the resistor has a resistance of 100 kΩ. The switch is in **position Y** and the capacitor is fully charged.

  (i) Calculate the maximum energy stored in the capacitor.   2

  (ii) The switch is moved to position X. Calculate the maximum current in the resistor.   2

                                        (10)

Marks

27. A car is fitted with an alarm which sounds a buzzer when the outside temperature falls below 3 °C.

The sensor is a thermistor located under the mirror on the side of the car.

The thermistor forms part of the circuit shown.

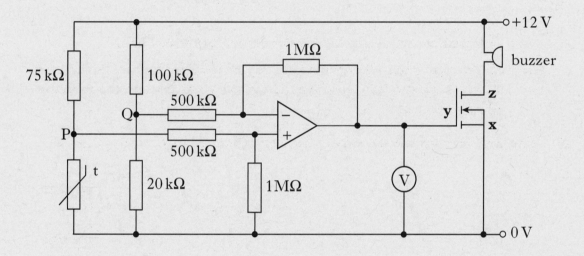

(a) What names are given to the terminals labelled **x**, **y** and **z** on the symbol for the MOSFET?

Clearly indicate which name goes with which letter. 1

(b) The buzzer sounds when the reading on the voltmeter is greater than or equal to +2·0 V.

(i) Calculate the minimum potential difference required between points P and Q to sound the buzzer. 2

(ii) Calculate the resistance of the thermistor when the reading on the voltmeter is +2·0 V. 2

(5)

[Turn over

Marks

28. A physics student investigates what happens when monochromatic light passes through a glass prism or a grating.

(a) The apparatus for the first experiment is shown below.

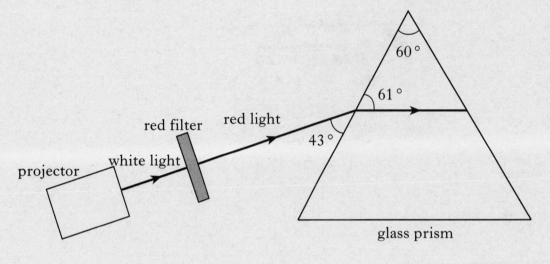

(i) Calculate the refractive index of the glass for the red light. **2**

(ii) Sketch a diagram which shows the ray of red light before, during and after passing through the prism. Mark on your diagram the values of all relevant angles. **2**

(b) The apparatus for the second experiment is shown below.

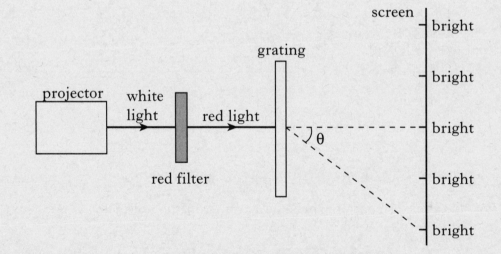

A pattern of bright and dark fringes is observed on the screen.

The grating has 300 lines per millimetre and the wavelength of the red light is 650 nm.

(i) Explain how the bright fringes are produced. **1**

(ii) Calculate the angle θ of the second order maximum. **2**

(iii) The red filter is replaced by a blue filter. Describe the effect of this change on the pattern observed.
Justify your answer. **1**

(8)

Marks

29. In 1902, P. Lenard set up an experiment similar to the one shown below.

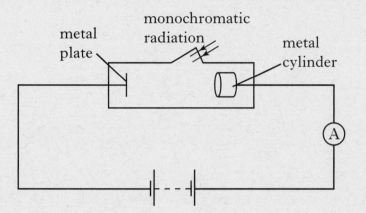

There is a constant potential difference between the metal plate and the metal cylinder.

Monochromatic radiation is directed onto the plate.

Photoelectrons produced at the plate are collected by the cylinder.

The frequency and the intensity of the radiation can be altered independently.

The frequency of the radiation is set at a value above the threshold frequency.

(a) The intensity of the radiation is slowly increased.

Sketch a graph of the current against intensity of radiation. 1

(b) The metal of the plate has a work function of $3 \cdot 11 \times 10^{-19}$ J. The wavelength of the radiation is 400 nm.

  (i) Calculate the maximum kinetic energy of a photoelectron. 3

  (ii) The battery connections are now reversed.

    Explain why there could still be a reading on the ammeter. 1

                              (5)

[Turn over

30. The nuclear industry must meet health and safety standards for workers. A worker has to handle radioactive materials behind a screen.

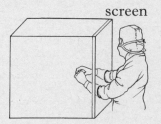

(*a*) The screen must be sufficiently thick to reduce the radiation to an acceptable level.

Different thicknesses of the screen material are placed between the source and the Geiger-Muller tube.

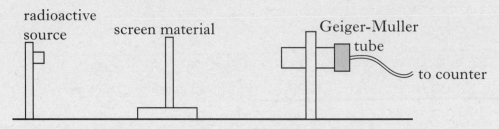

The graph shows corrected count rate plotted against thickness of material.

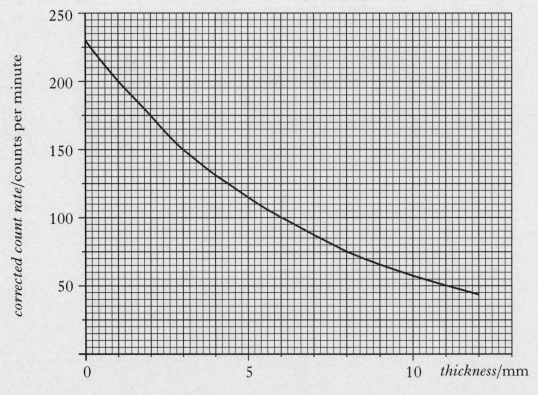

(i) Determine the half-value thickness of the material. **1**

(ii) The dose equivalent rate in air a short distance from this source is $20\,\mu\text{Sv}\,\text{h}^{-1}$.

When a certain thickness of the material is placed in front of the source, the dose equivalent rate at the same distance falls to $2 \cdot 5\,\mu\text{Sv}\,\text{h}^{-1}$.

Calculate the thickness of the material. **2**

Marks

30. (continued)

(b) The recommended dose equivalent limit for exposure to the hands of a worker is 500 mSv per year.

On average the worker is exposed to 2·0 mGy of gamma radiation, 400 μGy of thermal neutrons and 80 μGy of fast neutrons each hour when working in this area.

The quality factors for these radiations are shown.

Radiation	Quality factor
gamma	1
thermal neutrons	3
fast neutrons	10

The recommended dose equivalent limit must not be exceeded.

Calculate the maximum number of working hours in one year permitted in this area.

2

(5)

[END OF QUESTION PAPER]

[BLANK PAGE]

[BLANK PAGE]

X069/301

NATIONAL
QUALIFICATIONS
2006

WEDNESDAY, 17 MAY
1.00 PM – 3.30 PM

PHYSICS
HIGHER

Read Carefully

Reference may be made to the Physics Data Booklet.

1 All questions should be attempted.

Section A (questions 1 to 20)

2 Check that the answer sheet is for Physics Higher (Section A).

3 For this section of the examination you must use an **HB pencil** and, where necessary, an eraser.

4 Check that the answer sheet you have been given has **your name**, **date of birth**, **SCN** (Scottish Candidate Number) and **Centre Name** printed on it.

Do not change any of these details.

5 If any of this information is wrong, tell the Invigilator immediately.

6 If this information is correct, **print** your name and seat number in the boxes provided.

7 There is **only one correct** answer to each question.

8 Any rough working should be done on the question paper or the rough working sheet, **not** on your answer sheet.

9 At the end of the exam, put the **answer sheet for Section A inside the front cover of your answer book**.

10 Instructions as to how to record your answers to questions 1–20 are given on page three.

Section B (questions 21 to 29)

11 Answer the questions numbered 21 to 29 in the answer book provided.

12 **All answers must be written clearly and legibly in ink**.

13 Fill in the details on the front of the answer book.

14 Enter the question number clearly in the margin of the answer book beside each of your answers to questions 21 to 29.

15 Care should be taken to give an appropriate number of significant figures in the final answers to calculations.

SCOTTISH
QUALIFICATIONS
AUTHORITY

DATA SHEET

COMMON PHYSICAL QUANTITIES

Quantity	Symbol	Value	Quantity	Symbol	Value
Speed of light in vacuum	c	3.00×10^8 m s^{-1}	Mass of electron	m_e	9.11×10^{-31} kg
Magnitude of the charge on an electron	e	1.60×10^{-19} C	Mass of neutron	m_n	1.675×10^{-27} kg
Gravitational acceleration on Earth	g	9.8 m s^{-2}	Mass of proton	m_p	1.673×10^{-27} kg
Planck's constant	h	6.63×10^{-34} J s			

REFRACTIVE INDICES

The refractive indices refer to sodium light of wavelength 589 nm and to substances at a temperature of 273 K.

Substance	Refractive index	Substance	Refractive index
Diamond	2·42	Water	1·33
Crown glass	1·50	Air	1·00

SPECTRAL LINES

Element	Wavelength/nm	Colour	Element	Wavelength/nm	Colour
Hydrogen	656	Red	Cadmium	644	Red
	486	Blue-green		509	Green
	434	Blue-violet		480	Blue
	410	Violet			
	397	Ultraviolet			
	389	Ultraviolet			
Sodium	589	Yellow			

	Lasers	
Element	Wavelength/nm	Colour
Carbon dioxide	9550 10590 }	Infrared
Helium-neon	633	Red

PROPERTIES OF SELECTED MATERIALS

Substance	Density/ kg m^{-3}	Melting Point/ K	Boiling Point/ K
Aluminium	2.70×10^3	933	2623
Copper	8.96×10^3	1357	2853
Ice	9.20×10^2	273	. . . .
Sea Water	1.02×10^3	264	377
Water	1.00×10^3	273	373
Air	1·29	. . . .	. . . .
Hydrogen	9.0×10^{-2}	14	20

The gas densities refer to a temperature of 273 K and a pressure of 1.01×10^5 Pa.

SECTION A

For questions 1 to 20 in this section of the paper the answer to each question is either A, B, C, D or E. Decide what your answer is, then, using your pencil, put a horizontal line in the space provided—see the example below.

EXAMPLE

The energy unit measured by the electricity meter in your home is the

A kilowatt-hour

B ampere

C watt

D coulomb

E volt.

The correct answer is **A**—kilowatt-hour. The answer **A** has been clearly marked in **pencil** with a horizontal line (see below).

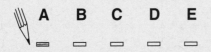

Changing an answer

If you decide to change your answer, carefully erase your first answer and, using your pencil, fill in the answer you want. The answer below has been changed to **E**.

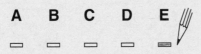

[Turn over

SECTION A

Answer questions 1–20 on the answer sheet.

1. Which of the following contains one scalar quantity and one vector quantity?

   A   acceleration; displacement

   B   kinetic energy; speed

   C   momentum; velocity

   D   potential energy; work

   E   power; weight

2. A golfer strikes a golf ball which then moves off at an angle to the ground. The ball follows the path shown.

   The graphs below show how the horizontal and vertical components of the velocity of the ball vary with time.

   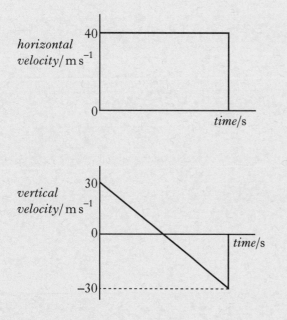

   What is the speed of the ball just before it hits the ground?

   A $10\,\mathrm{m\,s^{-1}}$

   B $30\,\mathrm{m\,s^{-1}}$

   C $40\,\mathrm{m\,s^{-1}}$

   D $50\,\mathrm{m\,s^{-1}}$

   E $70\,\mathrm{m\,s^{-1}}$

3. An object starts from rest and accelerates in a straight line.

   The graph shows how the acceleration of the object varies with time.

   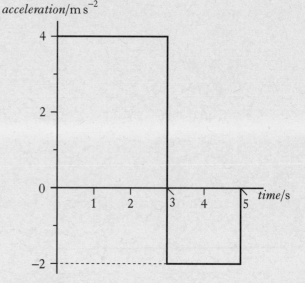

   The object's speed at 5 seconds is

   A $2\,\mathrm{m\,s^{-1}}$

   B $8\,\mathrm{m\,s^{-1}}$

   C $12\,\mathrm{m\,s^{-1}}$

   D $16\,\mathrm{m\,s^{-1}}$

   E $20\,\mathrm{m\,s^{-1}}$.

4. A person stands on bathroom scales in a lift. The scales show a reading greater than the person's weight.

   The lift is moving

   A   upwards at constant velocity

   B   downwards at constant velocity

   C   downwards and accelerating

   D   downwards and decelerating

   E   upwards and decelerating.

5. Two trolleys travel towards each other in a straight line as shown.

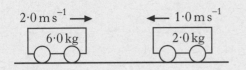

The trolleys collide. After the collision the trolleys move as shown below.

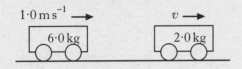

What is the speed v of the 2·0 kg trolley after the collision?

A $1·25\,\text{m s}^{-1}$

B $1·75\,\text{m s}^{-1}$

C $2·0\,\text{m s}^{-1}$

D $4·0\,\text{m s}^{-1}$

E $5·0\,\text{m s}^{-1}$

6. A student carries out an experiment to determine the density of a liquid.
The results are shown.

volume of liquid in beaker $= 2·00 \times 10^{-5}\,\text{m}^3$
mass of empty beaker $= 3·00 \times 10^{-2}\,\text{kg}$
mass of filled beaker $= 4·50 \times 10^{-2}\,\text{kg}$

The density of the liquid is

A $4·44 \times 10^{-4}\,\text{kg m}^{-3}$

B $1·33 \times 10^{-3}\,\text{kg m}^{-3}$

C $7·50 \times 10^{2}\,\text{kg m}^{-3}$

D $2·25 \times 10^{3}\,\text{kg m}^{-3}$

E $3·75 \times 10^{3}\,\text{kg m}^{-3}$.

7. Which pair of graphs shows how the pressure produced by a liquid depends on the depth and density of the liquid?

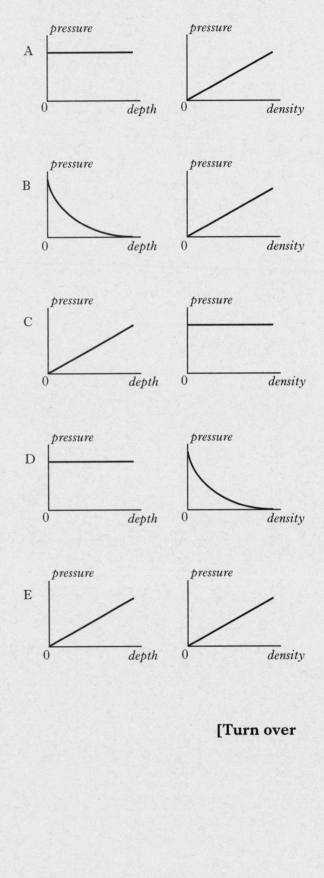

[Turn over

8. Three resistors are connected as shown.

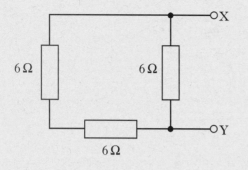

The total resistance between X and Y is

A $2\,\Omega$

B $4\,\Omega$

C $6\,\Omega$

D $9\,\Omega$

E $18\,\Omega$.

9. A battery of e.m.f. 12 V and internal resistance $3{\cdot}0\,\Omega$ is connected in a circuit as shown.

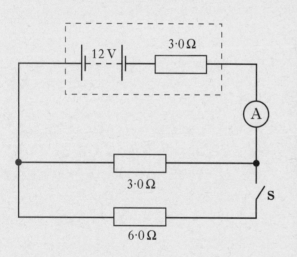

When switch **S** is closed the ammeter reading changes from

A $2{\cdot}0$ A to $1{\cdot}0$ A

B $2{\cdot}0$ A to $2{\cdot}4$ A

C $2{\cdot}0$ A to 10 A

D $4{\cdot}0$ A to $1{\cdot}3$ A

E $4{\cdot}0$ A to $6{\cdot}0$ A.

10. The circuit diagram shows a balanced Wheatstone bridge.

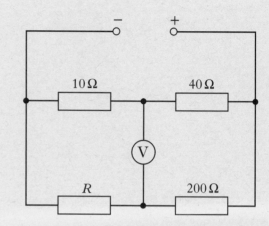

The resistance of resistor R is

A $0{\cdot}5\,\Omega$

B $2{\cdot}0\,\Omega$

C $50\,\Omega$

D $100\,\Omega$

E $800\,\Omega$.

11. A student carries out three experiments to investigate the charging of a capacitor using a d.c. supply.

The graphs obtained from the experiments are shown.

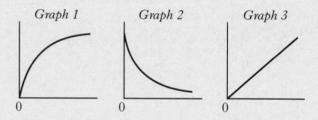

The axes of the graphs have not been labelled.

Which row in the table shows the labels for the axes of the graphs?

	Graph 1	Graph 2	Graph 3
A	voltage and time	current and time	charge and voltage
B	current and time	voltage and time	charge and voltage
C	current and time	charge and voltage	voltage and time
D	charge and voltage	current and time	voltage and time
E	voltage and time	charge and voltage	current and time

12. The following circuit shows a constant voltage a.c. supply connected to a resistor and capacitor in parallel.

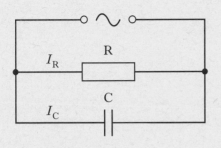

Which pair of graphs shows how the r.m.s. currents I_R and I_C vary as the frequency f of the supply is increased?

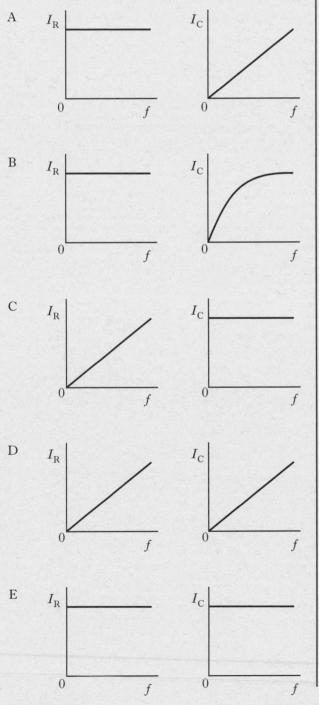

13. An op-amp circuit is set up as shown.

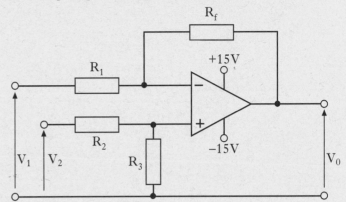

Which of the following statements is/are true?

I   The circuit is connected in the inverting mode.

II The circuit amplifies the difference between V_2 and V_1.

III The maximum possible output voltage $V_0 = V_1 + V_2$.

A   I only

B   II only

C   I and II only

D   II and III only

E   I, II and III

14. A microwave source at point O produces waves of wavelength 28 mm.

A metal reflector is placed as shown.

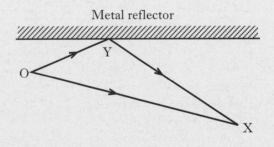

An interference pattern is produced.
Constructive interference occurs at point X.
The distance OX is 400 mm.
The total path length OYX is

A   414 mm

B   421 mm

C   442 mm

D   456 mm

E   463 mm.

15. The diagram represents a ray of light passing from air into liquid.

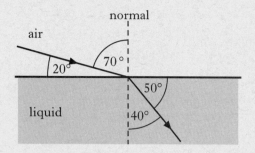

The refractive index of this liquid, relative to air, is

A $\dfrac{\sin 20°}{\sin 40°}$

B $\dfrac{\sin 40°}{\sin 70°}$

C $\dfrac{\sin 50°}{\sin 20°}$

D $\dfrac{\sin 70°}{\sin 40°}$

E $\dfrac{\sin 90°}{\sin 40°}.$

16. Light travels from air into glass.

Which row in the table describes what happens to the speed, frequency and wavelength of the light?

	Speed	Frequency	Wavelength
A	increases	stays constant	increases
B	increases	decreases	stays constant
C	stays constant	decreases	decreases
D	decreases	decreases	stays constant
E	decreases	stays constant	decreases

17. When light of frequency f is shone on to a certain metal, photoelectrons are ejected with a maximum velocity v and kinetic energy E_k.

Light of the same frequency but twice the irradiance is shone on to the same surface.

Which of the following statements is/are correct?

I   Twice as many electrons are ejected per second.

II   The speed of the fastest electron is $2v$.

III   The kinetic energy of the fastest electron is now $2E_k$.

A   I only

B   II only

C   III only

D   I and II only

E   I, II and III

18. The diagram shows some of the energy levels for the hydrogen atom.

E_3 _____ -1.360×10^{-19} J
E_2 _____ -2.416×10^{-19} J

E_1 _____ -5.424×10^{-19} J

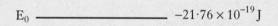

E_0 _____ -21.76×10^{-19} J

The highest frequency of radiation emitted due to a transition between two of these energy levels is

A 1.59×10^{14} Hz

B 2.46×10^{15} Hz

C 3.08×10^{15} Hz

D 1.63×10^{20} Hz

E 2.04×10^{20} Hz.

19. A series of radioactive decays starts from the isotope Uranium 238.

Two alpha particles and two beta particles are emitted during the decays.

Which row in the table gives the mass number and the atomic number of the resulting nucleus?

	Mass number	Atomic number
A	232	88
B	230	86
C	230	90
D	246	94
E	246	98

20. The table shows the radiation weighting factor w_R for a number of different radiations.

Type of radiation	Radiation weighting factor w_R
alpha particles	20
beta particles	1
neutrons	3
gamma rays	1
X-rays	0·1

Which of the following gives the greatest equivalent dose?

A   8 μGy of alpha particles

B   170 μGy of beta particles

C   56 μGy of neutrons

D   160 μGy of gamma rays

E   1500 μGy of X-rays

SECTION B begins on *Page eleven*

SECTION B

Write your answers to questions 21 to 29 in the answer book.

Marks

21. A van of mass 2600 kg moves down a slope which is inclined at 12° to the horizontal as shown.

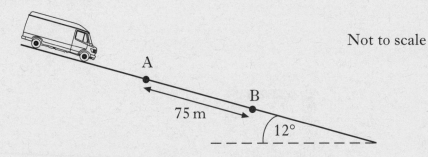

Not to scale

75 m

B

A

12°

(a) Calculate the component of the van's weight parallel to the slope.   2

(b) A constant frictional force of 1400 N acts on the van as it moves down the slope.

Calculate the acceleration of the van.   2

(c) The speed of the van as it passes point **A** is 5·0 m s^{-1}.
Point **B** is 75 m further down the slope.

Calculate the kinetic energy of the van at **B**.   3

(7)

[Turn over

Marks

22. A force sensor is used to investigate the impact of a ball as it bounces on a flat horizontal surface. The ball has a mass of 0·050 kg and is dropped vertically, from rest, through a height of 1·6 m as shown.

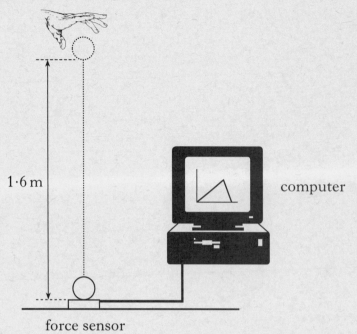

(a) The graph shows how the force on the ball varies with time during the impact.

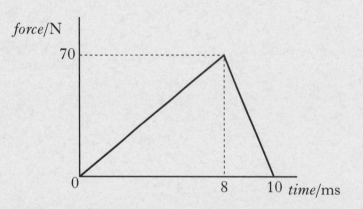

   (i) Show by calculation that the magnitude of the impulse on the ball is 0·35 N s.   **1**

   (ii) What is the magnitude and direction of the change in momentum of the ball?   **1**

   (iii) The ball is travelling at 5·6 m s^{-1} just before it hits the force sensor. Calculate the speed of the ball just as it leaves the force sensor.   **2**

(b) Another ball of identical size and mass, but made of a harder material, is dropped from rest and from the same height onto the same force sensor. Sketch the force-time graph shown above and, on the same axes, sketch another graph to show how the force on the harder ball varies with time.

Numerical values are not required but you must label the graphs clearly.   **2**

(6)

Marks

23. A refrigerated cool box is being prepared to carry medical supplies in a hot country. The **internal** dimensions of the box are $0.30\,\text{m} \times 0.20\,\text{m} \times 0.50\,\text{m}$.

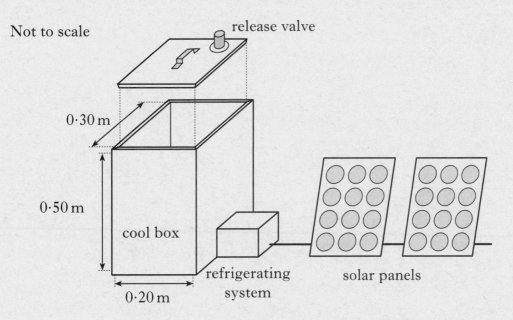

The lid is placed on the cool box with the release valve closed. An airtight seal is formed. When the lid is closed the air inside the cool box is at a temperature of $33\,^\circ\text{C}$ and a pressure of $1.01 \times 10^5\,\text{Pa}$.

The refrigerating system then reduces the temperature of the air inside the cool box until it reaches its working temperature.

At this temperature the air inside is at a pressure of $9.05 \times 10^4\,\text{Pa}$.

(a) (i) Calculate the temperature of the air inside the cool box when it is at its working temperature.   **2**

    (ii) Describe, using the kinetic model, how the decrease in temperature affects the air pressure inside the cool box.   **2**

(b) (i) Atmospheric pressure is $1.01 \times 10^5\,\text{Pa}$.

       Show that the magnitude of the force on the lid due to the difference in air pressure between the inside and outside of the cool box is now $630\,\text{N}$.   **2**

    (ii) The mass of the lid is $1.50\,\text{kg}$.

       Calculate the minimum force required to lift off the lid when the cool box is at its working temperature.   **1**

    (iii) The release valve allows air to pass into or out of the cool box.

       Explain why this valve should be opened before lifting the lid.   **1**

(c) The refrigerating system requires an average current of $0.80\,\text{A}$ at $12\,\text{V}$.

Each solar panel has a power output of $3.4\,\text{W}$ at $12\,\text{V}$.

Calculate the minimum number of solar panels needed to operate the refrigerating system.   **2**

(10)

Marks

24. The diagram below shows the basic features of a proton accelerator. It is enclosed in an evacuated container.

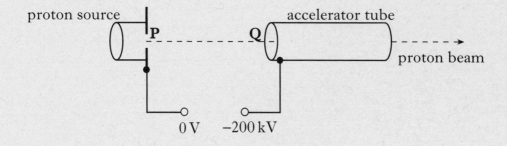

Protons released from the proton source start from rest at **P**.
A potential difference of 200 kV is maintained between **P** and **Q**.

(a) What is meant by the term *potential difference of 200 kV*? **1**

(b) Explain why protons released at **P** are accelerated towards **Q**. **1**

(c) Calculate:

    (i) the work done on a proton as it accelerates from **P** to **Q**; **2**

    (ii) the speed of a proton as it reaches **Q**. **2**

(d) The distance between **P** and **Q** is now halved.

    What effect, if any, does this change have on the speed of a proton as it reaches **Q**? Justify your answer. **2**

(8)

Marks

25. The 9·0 V battery in the circuit shown below has negligible internal resistance.

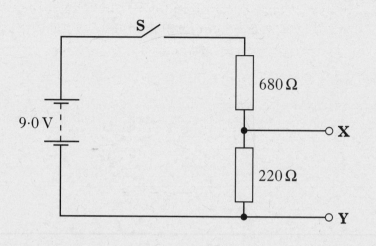

(a) Switch **S** is closed.

Calculate the potential difference between **X** and **Y**.

2

(b) Switch **S** is opened.

An uncharged 33 μF capacitor is connected between **X** and **Y** as shown.

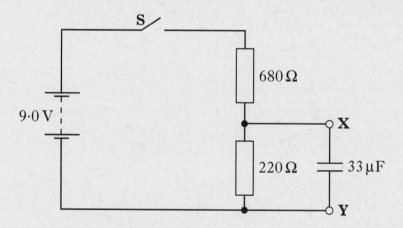

Switch **S** is then closed.

(i) Explain why work is done in charging the capacitor. 1

(ii) State the value of the maximum potential difference across the capacitor in this circuit. 1

(iii) Calculate the maximum energy stored in the capacitor. 2

(iv) Switch **S** is now opened.

Sketch a graph to show how the current through the 220 Ω resistor varies with time from the moment the switch is opened.

Numerical values are required only on the current axis. 2

(8)

[Turn over

Marks

26. A double beam oscilloscope has two inputs which allows two signals to be viewed on the screen at the same time.

    A double beam oscilloscope is connected to the input terminals **P** and **Q** and the output terminals **R** and **S** of a box containing an operational amplifier circuit.

    The operational amplifier is operating in the inverting mode.

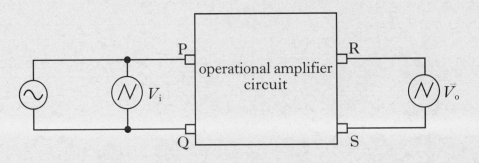

(a) The oscilloscope control settings and the two traces displayed on its screen are shown in the diagram.

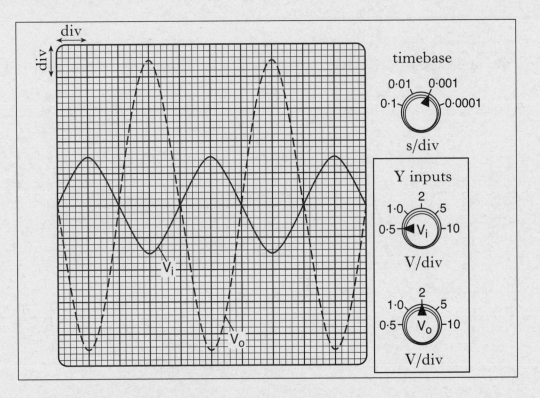

   (i) Calculate the frequency of the a.c. supply.   **2**

   (ii) Calculate the voltage gain of the amplifier circuit.   **2**

   (iii) Calculate the r.m.s. value of the output voltage of the amplifier circuit.   **2**

Marks

26. (continued)

(*b*) A student is given the task of altering the operational amplifier circuit inside the box to give a voltage gain of −4·7.

The following list shows resistor values available to the student.

Resistor value/kΩ
3·9
4·7
5·6
6·8
8·2
10
27
47
56

(i) Select suitable resistor values to produce a voltage gain of −4·7. **1**

(ii) Copy the diagram shown below.

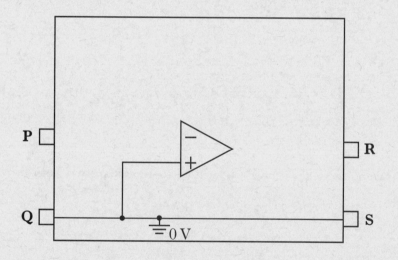

Complete the diagram, showing how your chosen resistors should be connected inside the box to complete the circuit. **1**

(8)

[Turn over

Marks

27. (*a*) Light of frequency $6 \cdot 7 \leftrightarrow 10^{14}$ Hz is produced at the junction of a light emitting diode (LED).

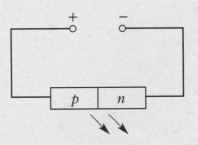

(i) Describe how the movement of charges in a forward-biased LED produces light. Your description should include the terms: *electrons; holes; photons* and *junction*.

1

(ii) (A) Calculate the wavelength of the light emitted from the LED.

2

(B) Use information from the data sheet on *Page two* to deduce the colour of this light.

1

(iii) The table below gives the values of the work function for three metals.

Metal	Work function/ J
caesium	$3 \cdot 4 \times 10^{-19}$
strontium	$4 \cdot 1 \times 10^{-19}$
magnesium	$5 \cdot 9 \times 10^{-19}$

Light from the LED is now incident on these metals in turn.

Show by calculation which of these metals, if any, release(s) photoelectrons with this light.

3

(*b*) Light from a different LED is passed through a grating as shown below.

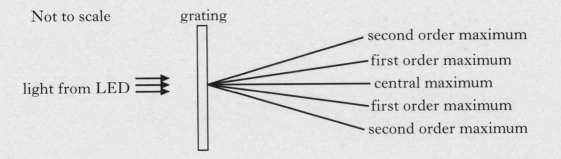

Light from this LED has a wavelength of $6 \cdot 35 \times 10^{-7}$ m. The spacing between lines in the grating is $5 \cdot 0 \times 10^{-6}$ m.

Calculate the angle between the central maximum and the **second** order maximum.

2

(9)

Marks

28. A student carries out an experiment to investigate how irradiance on a surface varies with distance from a small lamp.

Irradiance is measured with a light meter.

The distance between the small lamp and the light meter is measured with a metre stick.

The apparatus is set up as shown in a darkened laboratory.

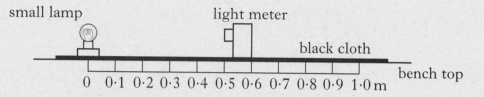

The following results are obtained.

Distance from source/ m	0·20	0·30	0·40	0·50
Irradiance/ units	675	302	170	108

(a) What is meant by the term *irradiance*?   **1**

(b) Use **all** the data to find the relationship between irradiance I and distance d from the source.   **2**

(c) What is the purpose of the black cloth on top of the bench?   **1**

(d) The small lamp is replaced by a laser.

Light from the laser is shone on to the light meter.

A reading is taken from the light meter when the distance between it and the laser is 0·50 m.

The distance is now increased to 1·00 m.

State how the new reading on the light meter compares with the one taken at 0·50 m.

Justify your answer.   **2**

    (6)

[Turn over

Marks

29. (a) About one hundred years ago Rutherford designed an experiment to investigate the structure of the atom. He used a radioactive source to fire alpha particles at a thin gold foil target.

His two assistants, Geiger and Marsden, spent many hours taking readings from the detector as it was moved to different positions between **X** and **Y**.

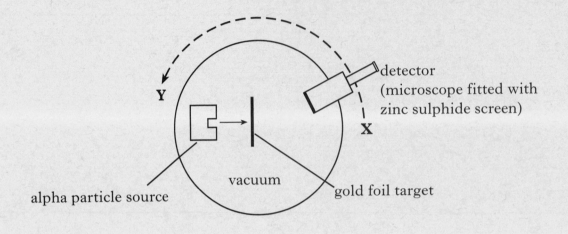

    (i) How did the number of alpha particles detected at **X** compare with the number detected at **Y**?   **1**

    (ii) State **one** conclusion Rutherford deduced from the results.   **1**

Marks

29. (continued)

(b) A nuclear fission reaction is represented by the following statement.

$$^{235}_{92}U \;+\; ^{1}_{0}n \;\rightarrow\; ^{137}_{r}Cs \;+\; ^{s}_{37}T \;+\; 4^{1}_{0}n$$

(i) Is this a spontaneous or an induced reaction? You must justify your answer. **1**

(ii) Determine the numbers represented by the letters *r* and *s* in the above reaction. **1**

(iii) Use the data booklet to identify the element represented by *T*. **1**

(iv) The masses of the nuclei and particles in the reaction are given below.

	Mass/kg
$^{235}_{92}U$	$390 \cdot 219 \times 10^{-27}$
$^{137}_{r}Cs$	$227 \cdot 292 \times 10^{-27}$
$^{s}_{37}T$	$157 \cdot 562 \times 10^{-27}$
$^{1}_{0}n$	$1 \cdot 675 \times 10^{-27}$

Calculate the energy released in the reaction. **3**

(8)

[END OF QUESTION PAPER]

[BLANK PAGE]

[BLANK PAGE]

X069/301

NATIONAL
QUALIFICATIONS
2007

WEDNESDAY, 16 MAY
1.00 PM – 3.30 PM

PHYSICS
HIGHER

Read Carefully

Reference may be made to the Physics Data Booklet.

1   All questions should be attempted.

Section A (questions 1 to 20)

2   Check that the answer sheet is for Physics Higher (Section A).

3   For this section of the examination you must use an **HB pencil** and, where necessary, an eraser.

4   Check that the answer sheet you have been given has **your name**, **date of birth**, **SCN** (Scottish Candidate Number) and **Centre Name** printed on it.

Do not change any of these details.

5   If any of this information is wrong, tell the Invigilator immediately.

6   If this information is correct, **print** your name and seat number in the boxes provided.

7   There is **only one correct** answer to each question.

8   Any rough working should be done on the question paper or the rough working sheet, **not** on your answer sheet.

9   At the end of the exam, put the **answer sheet for Section A inside the front cover of your answer book**.

10   Instructions as to how to record your answers to questions 1–20 are given on page three.

Section B (questions 21 to 31)

11   Answer the questions numbered 21 to 31 in the answer book provided.

12   **All answers must be written clearly and legibly in ink**.

13   Fill in the details on the front of the answer book.

14   Enter the question number clearly in the margin of the answer book beside each of your answers to questions 21 to 31.

15   Care should be taken to give an appropriate number of significant figures in the final answers to calculations.

16   Where additional paper, eg square ruled paper, is used, write your name and SCN (Scottish Candidate Number) on it and place it inside the front cover of your answer booklet.

SCOTTISH
QUALIFICATIONS
AUTHORITY

DATA SHEET
COMMON PHYSICAL QUANTITIES

Quantity	Symbol	Value	Quantity	Symbol	Value
Speed of light in vacuum	c	$3{\cdot}00 \times 10^8$ m s^{-1}	Mass of electron	m_e	$9{\cdot}11 \times 10^{-31}$ kg
Magnitude of the charge on an electron	e	$1{\cdot}60 \times 10^{-19}$ C	Mass of neutron	m_n	$1{\cdot}675 \times 10^{-27}$ kg
Gravitational acceleration on Earth	g	$9{\cdot}8$ m s^{-2}	Mass of proton	m_p	$1{\cdot}673 \times 10^{-27}$ kg
Planck's constant	h	$6{\cdot}63 \times 10^{-34}$ J s			

REFRACTIVE INDICES

The refractive indices refer to sodium light of wavelength 589 nm and to substances at a temperature of 273 K.

Substance	Refractive index	Substance	Refractive index
Diamond	2·42	Water	1·33
Crown glass	1·50	Air	1·00

SPECTRAL LINES

Element	Wavelength/nm	Colour	Element	Wavelength/nm	Colour
Hydrogen	656	Red	Cadmium	644	Red
	486	Blue-green		509	Green
	434	Blue-violet		480	Blue
	410	Violet		Lasers	
	397	Ultraviolet	Element	Wavelength/nm	Colour
	389	Ultraviolet	Carbon dioxide	9550 } 10590 }	Infrared
Sodium	589	Yellow	Helium-neon	633	Red

PROPERTIES OF SELECTED MATERIALS

Substance	Density/ kg m^{-3}	Melting Point/ K	Boiling Point/ K
Aluminium	$2{\cdot}70 \times 10^3$	933	2623
Copper	$8{\cdot}96 \times 10^3$	1357	2853
Ice	$9{\cdot}20 \times 10^2$	273	. . . .
Sea Water	$1{\cdot}02 \times 10^3$	264	377
Water	$1{\cdot}00 \times 10^3$	273	373
Air	1·29	. . . .	. . . .
Hydrogen	$9{\cdot}0 \times 10^{-2}$	14	20

The gas densities refer to a temperature of 273 K and a pressure of $1{\cdot}01 \times 10^5$ Pa.

SECTION A

For questions 1 to 20 in this section of the paper the answer to each question is either A, B, C, D or E. Decide what your answer is, then, using your pencil, put a horizontal line in the space provided—see the example below.

EXAMPLE

The energy unit measured by the electricity meter in your home is the

    A kilowatt-hour

    B ampere

    C watt

    D coulomb

    E volt.

The correct answer is **A**—kilowatt-hour. The answer **A** has been clearly marked in **pencil** with a horizontal line (see below).

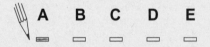

Changing an answer

If you decide to change your answer, carefully erase your first answer and, using your pencil, fill in the answer you want. The answer below has been changed to **E**.

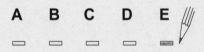

[Turn over

SECTION A

Answer questions 1–20 on the answer sheet.

1. Which row shows both quantities classified correctly?

	Scalar	*Vector*
A	weight	force
B	force	mass
C	mass	distance
D	distance	momentum
E	momentum	time

2. A ball is thrown vertically upwards and falls back to Earth. Neglecting air resistance, which velocity-time graph represents its motion?

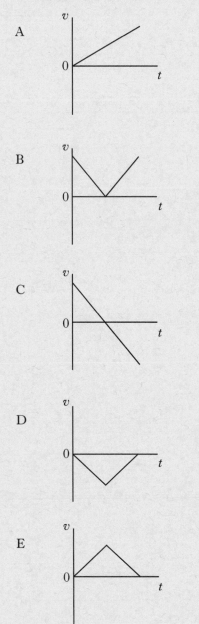

3. A person stands on a weighing machine in a lift. When the lift is at rest, the reading on the machine is 700 N. The lift now descends and its speed increases at a constant rate. The reading on the machine

A   is a constant value higher than 700 N

B   is a constant value lower than 700 N

C   continually increases from 700 N

D   continually decreases from 700 N

E   remains constant at 700 N.

4. Momentum can be measured in

A $N\,kg^{-1}$

B $N\,m$

C $N\,m\,s^{-1}$

D $kg\,m\,s^{-1}$

E $kg\,m\,s^{-2}$.

5. A cannon of mass 2000 kg fires a cannonball of mass 5·00 kg.

The cannonball leaves the cannon with a speed of $50\cdot0\,m\,s^{-1}$.

The speed of the cannon immediately after firing is

A $0\cdot125\,m\,s^{-1}$

B $8\cdot00\,m\,s^{-1}$

C $39\cdot9\,m\,s^{-1}$

D $40\cdot1\,m\,s^{-1}$

E $200\,m\,s^{-1}$.

6. The graph shows the force acting on an object of mass $5 \cdot 0$ kg.

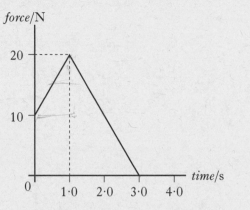

The change in the object's momentum is

A $7 \cdot 0 \, \text{kg m s}^{-1}$

B $30 \, \text{kg m s}^{-1}$

C $35 \, \text{kg m s}^{-1}$

D $60 \, \text{kg m s}^{-1}$

E $175 \, \text{kg m s}^{-1}$.

7. Which of the following gives the approximate relative spacings of molecules in ice, water and water vapour?

	Spacing of molecules in ice	Spacing of molecules in water	Spacing of molecules in water vapour
A	1	1	10
B	1	3	1
C	1	3	3
D	1	10	10
E	3	1	10

8. The element of an electric kettle has a resistance of 30 Ω. The kettle is connected to a mains supply. The r.m.s. voltage of this supply is 230 V. The peak value of the current in the kettle is

A $0 \cdot 13 \, \text{A}$

B $0 \cdot 18 \, \text{A}$

C $5 \cdot 4 \, \text{A}$

D $7 \cdot 7 \, \text{A}$

E $10 \cdot 8 \, \text{A}$.

9. Four resistors, each of resistance 20 Ω, are connected to a 60 V supply of negligible internal resistance, as shown.

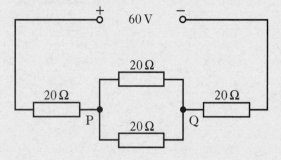

The potential difference across PQ is

A   12 V

B   15 V

C   20 V

D   24 V

E   30 V.

10. A signal from a power supply is displayed on an oscilloscope.

The trace on the oscilloscope is shown.

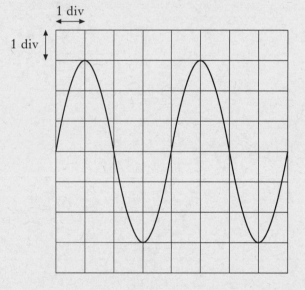

The time-base is set at $0 \cdot 01$ s/div and the Y-gain is set at $4 \cdot 0$ V/div.

Which row in the table shows the r.m.s. voltage and the frequency of the signal?

	r.m.s. voltage/V	frequency/Hz
A	$8 \cdot 5$	25
B	12	25
C	24	25
D	$8 \cdot 5$	50
E	12	50

11. A resistor and an ammeter are connected to a signal generator which has an output of constant amplitude and variable frequency.

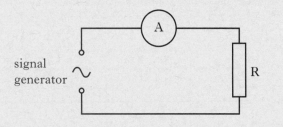

Which graph shows the relationship between the current I in the resistor and the output frequency f of the signal generator?

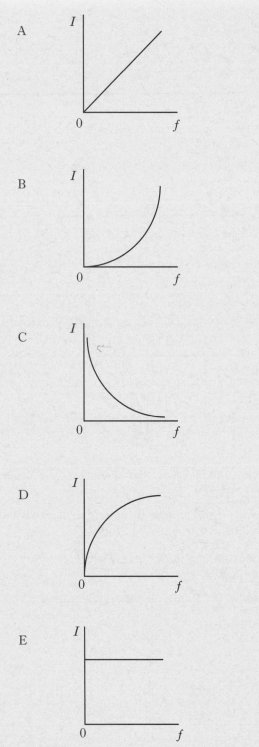

12. An oscilloscope is used to measure the frequency of the output voltage from an op-amp.

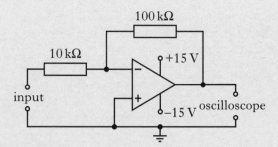

The input has a frequency of 280 Hz and a peak voltage of 0·5 V.

The frequency of the output voltage is

A   28 Hz

B   140 Hz

C   280 Hz

D   560 Hz

E   2800 Hz.

13. An op-amp circuit is set up as shown.

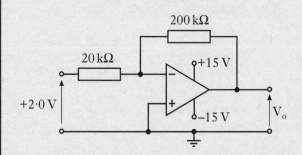

A voltage of +2·0 V is applied to the input. The voltage output, V_o, is approximately

A   +20 V

B   +15 V

C   −2·0 V

D   −15 V

E   −20 V.

14. The energy of a wave depends on its

    A   amplitude

    B   period

    C   phase

    D   speed

    E   wavelength.

15. A ray of light travels from air into a glass prism. The refractive index of the glass is 1·50.

Which diagram shows the correct path of the ray?

A

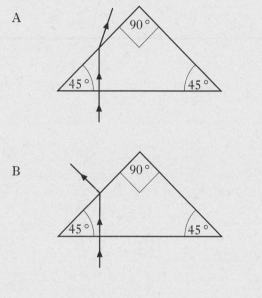

B

C

D

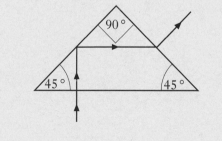

E

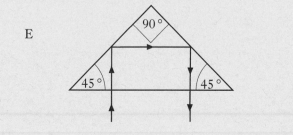

16. A beam of white light is passed through two optical components P and Q. Component P produces a number of spectra and component Q produces a spectrum as shown.

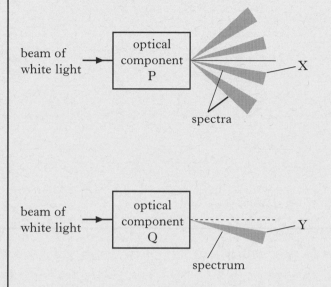

Which row in the table identifies the optical components and the colour of light seen at position X and position Y?

	Optical component P	Colour seen at X	Optical component Q	Colour seen at Y
A	grating	red	triangular prism	red
B	grating	red	triangular prism	violet
C	grating	violet	triangular prism	red
D	triangular prism	red	grating	violet
E	triangular prism	violet	grating	red

[Turn over

17. The diagram represents some electron transitions between energy levels in an atom.

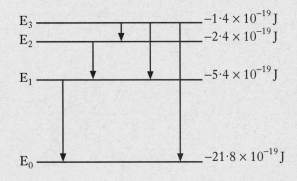

E_3 —————— $-1\cdot4 \times 10^{-19}$ J
E_2 —————— $-2\cdot4 \times 10^{-19}$ J

E_1 —————— $-5\cdot4 \times 10^{-19}$ J

E_0 —————— $-21\cdot8 \times 10^{-19}$ J

The radiation emitted with the shortest wavelength is produced by an electron making transition

A E_1 to E_0

B E_2 to E_1

C E_3 to E_2

D E_3 to E_1

E E_3 to E_0.

18. In the following circuit, component X is used to drive a motor.

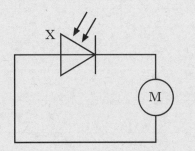

Which of the following gives the name of component X and its mode of operation?

	Name of component X	Mode of operation
A	light-emitting diode	photoconductive
B	light-emitting diode	photovoltaic
C	photodiode	photoconductive
D	photodiode	photovoltaic
E	op-amp	inverting

19. The classical experiment on the scattering of alpha particles from a thin gold foil suggested that

A   positive charges were evenly distributed throughout the atom

B   atomic nuclei were very small and positively charged

C   neutrons existed in the nucleus

D   alpha particles were helium nuclei

E   alpha particles were hydrogen nuclei.

20. A radioactive source produces a count rate of 2400 counts per second in a detector. When a lead plate of thickness 36 mm is placed between the source and the detector the count rate falls to 300 counts per second.

The half-value thickness of lead for this radiation is

A   4·5 mm

B   12 mm

C   36 mm

D   108 mm

E   288 mm.

[Turn over for SECTION B on *Page ten*

SECTION B

Write your answers to questions 21 to 31 in the answer book.

Marks

21. Competitors are racing remote control cars. The cars have to be driven over a precise route between checkpoints.

Checkpoint A

Each car is to travel from checkpoint A to checkpoint B by following these instructions.

"Drive 150 m due North, then drive 250 m on a bearing of 60° East of North (060)."

Car X takes 1 minute 6 seconds to follow these instructions exactly.

(a) By scale drawing or otherwise, find the displacement of checkpoint B from checkpoint A.   2

(b) Calculate the average velocity of car X from checkpoint A to checkpoint B.   2

(c) Car Y leaves A at the same time as car X.

Car Y follows exactly the same route at an average speed of $6 \cdot 5 \, \text{m s}^{-1}$.

Which car arrives first at checkpoint B?

Justify your answer with a calculation.   2

(d) State the displacement of checkpoint A from checkpoint B.   1

(7)

Marks

22. A fairground ride consists of rafts which slide down a slope into water.

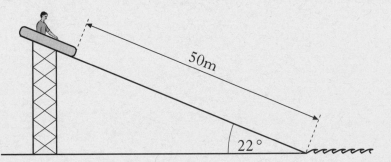

The slope is at an angle of 22° to the horizontal. Each raft has a mass of 8·0 kg. The length of the slope is 50 m.

A child of mass 52 kg sits in a raft at the top of the slope. The raft is released from rest. The child and raft slide together down the slope into the water. The force of friction between the raft and slope remains constant at 180 N.

(*a*) Calculate the component of weight, in newtons, of the child and raft down the slope.   **1**

(*b*) Show by calculation that the acceleration of the child and raft down the slope is $0·67 \text{ m s}^{-2}$.   **2**

(*c*) Calculate the speed of the child and raft at the bottom of the slope.   **2**

(*d*) A second child of smaller mass is released from rest in an identical raft at the same starting point. The force of friction is the same as before.

How does the speed of this child and raft at the bottom of the slope compare with the answer to part (*c*)?

Justify your answer.   **2**

(7)

[Turn over

Marks

23. A rigid cylinder contains $8{\cdot}0 \times 10^{-2}\,\text{m}^3$ of helium gas at a pressure of $750\,\text{kPa}$. Gas is released from the cylinder to fill party balloons.

party balloons

rigid cylinder

During the filling process, the temperature remains constant. When filled, each balloon holds $0{\cdot}020\,\text{m}^3$ of helium gas at a pressure of $125\,\text{kPa}$.

(a) Calculate the total volume of the helium gas when it is at a pressure of $125\,\text{kPa}$.

2

(b) Determine the maximum number of balloons which can be fully inflated by releasing gas from the cylinder.

2

(c) State how the density of the helium gas in an inflated balloon compares to the initial density of the helium gas inside the cylinder.

Justify your answer.

2

(6)

Marks

24. The apparatus shown in the diagram is designed to accelerate alpha particles.

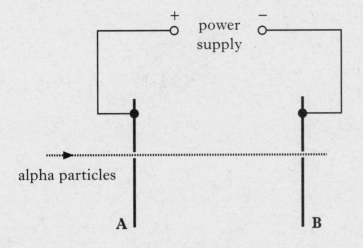

An alpha particle travelling at a speed of $2 \cdot 60 \times 10^6 \, \mathrm{m \, s^{-1}}$ passes through a hole in plate A. The mass of an alpha particle is $6 \cdot 64 \times 10^{-27} \, \mathrm{kg}$ and its charge is $3 \cdot 2 \times 10^{-19} \, \mathrm{C}$.

(*a*) When the alpha particle reaches plate B, its kinetic energy has increased to $3 \cdot 05 \times 10^{-14} \, \mathrm{J}$.

Show that the work done on the alpha particle as it moves from plate A to plate B is $8 \cdot 1 \times 10^{-15} \, \mathrm{J}$.   2

(*b*) Calculate the potential difference between plates A and B.   2

(*c*) The apparatus is now adapted to accelerate **electrons** from A to B through the same potential difference.

How does the increase in the kinetic energy of an electron compare with the increase in kinetic energy of the alpha particle in part (*a*)?

Justify your answer.   2

(6)

[Turn over

Marks

25. A power supply of e.m.f. E and internal resistance $2 \cdot 0\,\Omega$ is connected as shown.

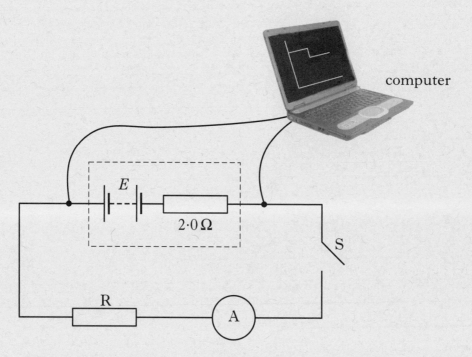

The computer connected to the apparatus displays a graph of potential difference against time.

The graph shows the potential difference across the terminals of the power supply for a short time before and after switch S is closed.

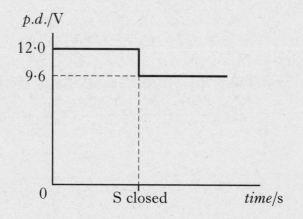

(a) State the e.m.f. of the power supply. **1**

(b) Calculate:

   (i) the reading on the ammeter after switch S is closed; **2**

   (ii) the resistance of resistor R. **1**

Marks

25.   (continued)

(c)   Switch S is opened. A second identical resistor is now connected in parallel with R as shown.

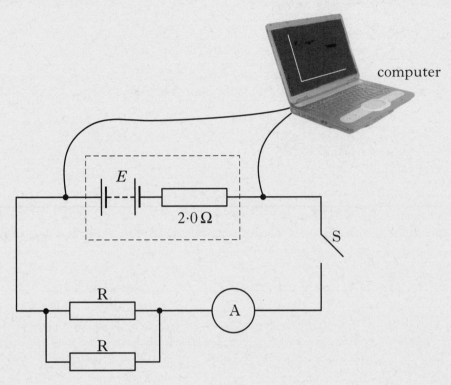

The computer is again connected in order to display a graph of potential difference against time.

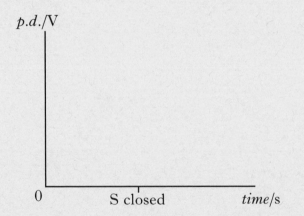

Copy and complete the new graph of potential difference against time showing the values of potential difference before and after switch S is closed.

2

(6)

[Turn over

Marks

26. An uncharged 2200 μF capacitor is connected in a circuit as shown.

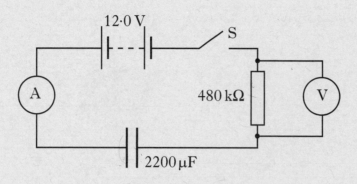

The battery has negligible internal resistance.

(a) Switch S is closed. Calculate the initial charging current.   **2**

(b) At one instant during the charging process the potential difference **across the resistor** is 3·8 V.

Calculate the charge stored in the capacitor at this instant.   **3**

(c) Calculate the **maximum** energy the capacitor stores in this circuit.   **2**

   (7)

Marks

27. A Wheatstone bridge is used to measure the resistance of a thermistor as its temperature changes.

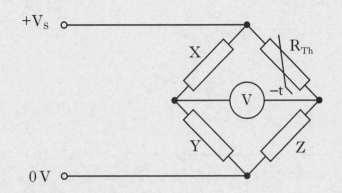

(a) The bridge is balanced when X = 2·2 kΩ, Y = 5·0 kΩ and Z = 750 Ω.

Calculate the resistance of the thermistor, R_{Th}, when the bridge is balanced.　　**2**

(b) A student uses this bridge in a circuit to light an LED when the temperature in a greenhouse falls below a certain level.

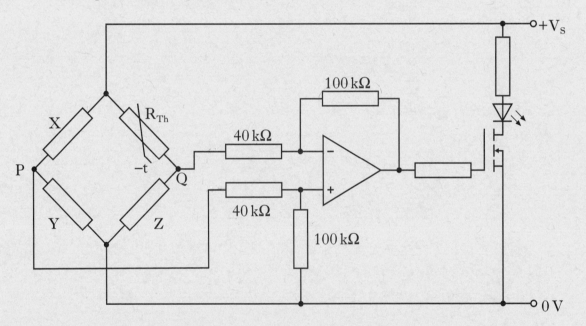

(i) In which mode is the op-amp being used?　　**1**

(ii) As the temperature of the thermistor falls, its resistance increases.

Explain how this whole circuit operates to cause the LED to light when the temperature falls.　　**2**

(iii) At a certain temperature the output voltage of the op-amp is 3·0 V.

Calculate the potential difference between P and Q at this temperature.　　**2**

(7)

[Turn over

Marks

28. An experiment to determine the wavelength of light from a laser is shown.

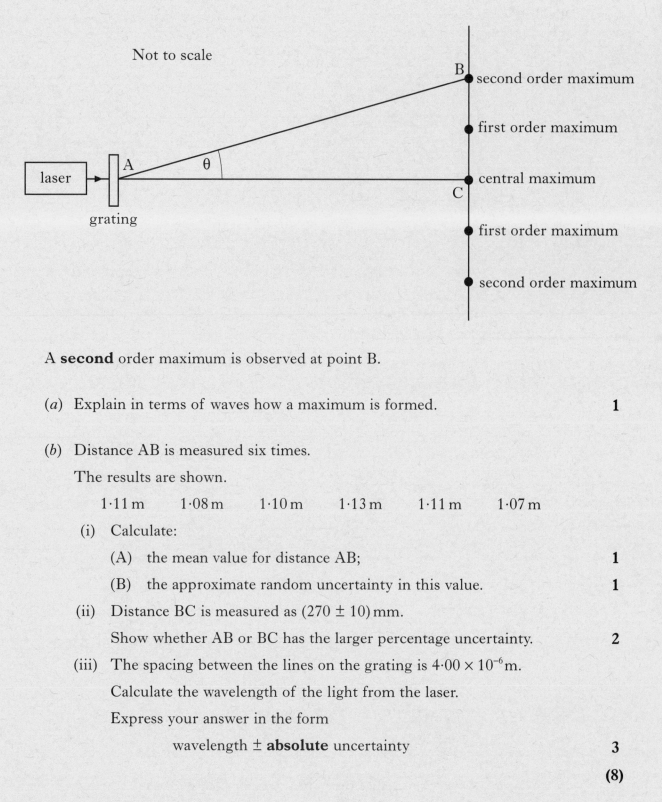

Not to scale

A **second** order maximum is observed at point B.

(*a*) Explain in terms of waves how a maximum is formed.　　　　1

(*b*) Distance AB is measured six times.

The results are shown.

    1·11 m   1·08 m   1·10 m   1·13 m   1·11 m   1·07 m

  (i) Calculate:

    (A)　the mean value for distance AB;　　　　1

    (B)　the approximate random uncertainty in this value.　　　　1

  (ii) Distance BC is measured as (270 ± 10) mm.

    Show whether AB or BC has the larger percentage uncertainty.　　　　2

  (iii) The spacing between the lines on the grating is $4 \cdot 00 \times 10^{-6}$ m.

    Calculate the wavelength of the light from the laser.

    Express your answer in the form

        wavelength ± **absolute** uncertainty　　　　3

                               (8)

Marks

29. A ray of red light is incident on a semicircular block of glass at the mid point of XY as shown.

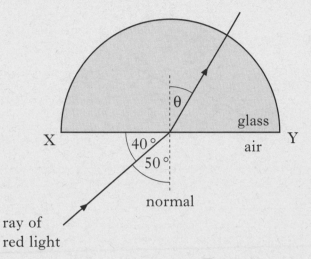

The refractive index of the block is 1·50 for this red light.

(a) Calculate angle θ shown on the diagram.

2

(b) The wavelength of the red light **in the glass** is 420 nm.

Calculate the wavelength of the light in air.

2

(c) The ray of red light is replaced by a ray of blue light incident at the same angle. The blue light enters the block at the same point.

Explain why the path taken by the blue light in the block is different to that taken by the red light.

1

(5)

[Turn over

Marks

30. A metal plate emits electrons when certain wavelengths of electromagnetic radiation are incident on it.

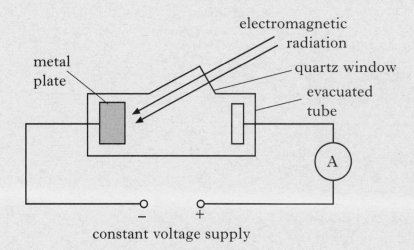

When light of wavelength 605 nm is incident on the metal plate, electrons are released with zero kinetic energy.

(*a*) Show that the work function of this metal is 3.29×10^{-19} J.　　2

(*b*) The wavelength of the incident radiation is now altered. Photons of energy 5.12×10^{-19} J are incident on the metal plate.

  (i) Calculate the maximum kinetic energy of the electrons just as they leave the metal plate.　　1

  (ii) The irradiance of this radiation on the metal plate is now decreased.

    State the effect this has on the ammeter reading.

    Justify your answer.　　2

                   (5)

Marks

31. (*a*) The following statement represents a nuclear reaction.

$$^{240}_{94}\text{Pu} \longrightarrow {}^{236}_{92}\text{U} + {}^{4}_{2}\text{He}$$

The table shows the masses of the particles involved in this reaction.

Particle	Mass/kg
$^{240}_{94}\text{Pu}$	$398 \cdot 626 \times 10^{-27}$
$^{236}_{92}\text{U}$	$391 \cdot 970 \times 10^{-27}$
$^{4}_{2}\text{He}$	$6 \cdot 645 \times 10^{-27}$

Calculate the energy released in this reaction. 3

(*b*) A technician is working with a radioactive source as shown.

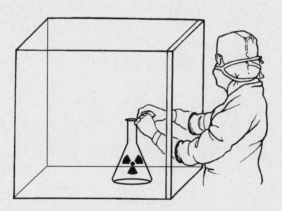

The technician's hands receive an absorbed dose at a rate of $4 \cdot 0 \, \mu\text{Gy h}^{-1}$ for 2 hours. The radiation from the source has a radiation weighting factor of 3. Calculate the equivalent dose received by the technician's hands. 3

(6)

[END OF QUESTION PAPER]

[BLANK PAGE]

[BLANK PAGE]

X069/301

NATIONAL	FRIDAY, 23 MAY	PHYSICS
QUALIFICATIONS	1.00 PM – 3.30 PM	HIGHER
2008		

Read Carefully

Reference may be made to the Physics Data Booklet.

 1 All questions should be attempted.

Section A (questions 1 to 20)

 2 Check that the answer sheet is for Physics Higher (Section A).

 3 For this section of the examination you must use an **HB pencil** and, where necessary, an eraser.

 4 Check that the answer sheet you have been given has **your name**, **date of birth**, **SCN** (Scottish Candidate Number) and **Centre Name** printed on it.

Do not change any of these details.

 5 If any of this information is wrong, tell the Invigilator immediately.

 6 If this information is correct, **print** your name and seat number in the boxes provided.

 7 There is **only one correct** answer to each question.

 8 Any rough working should be done on the question paper or the rough working sheet, **not** on your answer sheet.

 9 At the end of the exam, put the **answer sheet for Section A inside the front cover of your answer book**.

10 Instructions as to how to record your answers to questions 1–20 are given on page three.

Section B (questions 21 to 30)

11 Answer the questions numbered 21 to 30 in the answer book provided.

12 **All answers must be written clearly and legibly in ink**.

13 Fill in the details on the front of the answer book.

14 Enter the question number clearly in the margin of the answer book beside each of your answers to questions 21 to 30.

15 Care should be taken to give an appropriate number of significant figures in the final answers to calculations.

16 Where additional paper, eg square ruled paper, is used, write your name and SCN (Scottish Candidate Number) on it and place it inside the front cover of your answer booklet.

DATA SHEET
COMMON PHYSICAL QUANTITIES

Quantity	Symbol	Value	Quantity	Symbol	Value
Speed of light in vacuum	c	3.00×10^8 m s^{-1}	Mass of electron	m_e	9.11×10^{-31} kg
Magnitude of the charge on an electron	e	1.60×10^{-19} C	Mass of neutron	m_n	1.675×10^{-27} kg
Gravitational acceleration on Earth	g	9.8 m s^{-2}	Mass of proton	m_p	1.673×10^{-27} kg
Planck's constant	h	6.63×10^{-34} J s			

REFRACTIVE INDICES

The refractive indices refer to sodium light of wavelength 589 nm and to substances at a temperature of 273 K.

Substance	Refractive index	Substance	Refractive index
Diamond	2·42	Water	1·33
Crown glass	1·50	Air	1·00

SPECTRAL LINES

Element	Wavelength/nm	Colour	Element	Wavelength/nm	Colour
Hydrogen	656	Red	Cadmium	644	Red
	486	Blue-green		509	Green
	434	Blue-violet		480	Blue
	410	Violet		*Lasers*	
	397	Ultraviolet	Element	Wavelength/nm	Colour
	389	Ultraviolet	Carbon dioxide	9550 \} 10590	Infrared
Sodium	589	Yellow	Helium-neon	633	Red

PROPERTIES OF SELECTED MATERIALS

Substance	Density/ kg m^{-3}	Melting Point/ K	Boiling Point/ K
Aluminium	2.70×10^3	933	2623
Copper	8.96×10^3	1357	2853
Ice	9.20×10^2	273	. . . .
Sea Water	1.02×10^3	264	377
Water	1.00×10^3	273	373
Air	1·29	. . . .	. . . .
Hydrogen	$9.0 \ \times 10^{-2}$	14	20

The gas densities refer to a temperature of 273 K and a pressure of 1.01×10^5 Pa.

SECTION A

For questions 1 to 20 in this section of the paper the answer to each question is either A, B, C, D or E. Decide what your answer is, then, using your pencil, put a horizontal line in the space provided—see the example below.

EXAMPLE

The energy unit measured by the electricity meter in your home is the

    A kilowatt-hour

    B ampere

    C watt

    D coulomb

    E volt.

The correct answer is **A**—kilowatt-hour. The answer **A** has been clearly marked in **pencil** with a horizontal line (see below).

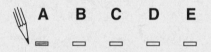

Changing an answer

If you decide to change your answer, carefully erase your first answer and, using your pencil, fill in the answer you want. The answer below has been changed to **E**.

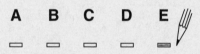

[Turn over

SECTION A

Answer questions 1–20 on the answer sheet.

1. Which row in the table is correct?

	Scalar	Vector
A	distance	work
B	weight	acceleration
C	velocity	displacement
D	mass	momentum
E	speed	time

2. A javelin is thrown at $60°$ to the horizontal with a speed of 20 m s^{-1}.

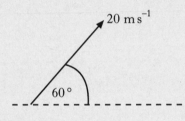

The javelin is in flight for 3.5 s.
Air resistance is negligible.
The horizontal distance the javelin travels is

A 35.0 m

B 60.6 m

C 70.0 m

D 121 m

E 140 m.

3. Two boxes on a frictionless horizontal surface are joined together by a string. A constant horizontal force of 12 N is applied as shown.

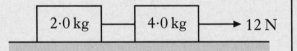

The tension in the string joining the two boxes is

A 2.0 N

B 4.0 N

C 6.0 N

D 8.0 N

E 12 N.

4. The total mass of a motorcycle and rider is 250 kg. During braking, they are brought to rest from a speed of 16.0 m s^{-1} in a time of 10.0 s.

The maximum energy which could be converted to heat in the brakes is

A 2000 J

B 4000 J

C $32\,000 \text{ J}$

D $40\,000 \text{ J}$

E $64\,000 \text{ J}$.

5. A shell of mass 5.0 kg is travelling horizontally with a speed of 200 m s^{-1}. It explodes into two parts. One part of mass 3.0 kg continues in the original direction with a speed of 100 m s^{-1}.

The other part also continues in this same direction. Its speed is

A 150 m s^{-1}

B 200 m s^{-1}

C 300 m s^{-1}

D 350 m s^{-1}

E 700 m s^{-1}.

6. The graph shows the force which acts on an object over a time interval of 8 seconds.

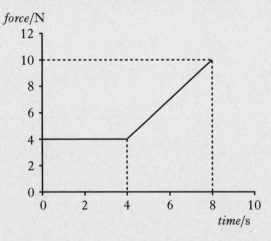

The momentum gained by the object during this 8 seconds is

A 12 kg m s^{-1}

B 32 kg m s^{-1}

C 44 kg m s^{-1}

D 52 kg m s^{-1}

E 72 kg m s^{-1}.

7. One pascal is equivalent to

A $1 \, N \, m$

B $1 \, N \, m^2$

C $1 \, N \, m^3$

D $1 \, N \, m^{-2}$

E $1 \, N \, m^{-3}$.

8. An electron is accelerated from rest through a potential difference of $2 \cdot 0 \, kV$.

The kinetic energy gained by the electron is

A $8 \cdot 0 \times 10^{-23} \, J$

B $8 \cdot 0 \times 10^{-20} \, J$

C $3 \cdot 2 \times 10^{-19} \, J$

D $1 \cdot 6 \times 10^{-16} \, J$

E $3 \cdot 2 \times 10^{-16} \, J$.

9. The e.m.f. of a battery is

A   the total energy supplied by the battery

B   the voltage lost due to the internal resistance of the battery

C   the total charge which passes through the battery

D   the number of coulombs of charge passing through the battery per second

E   the energy supplied to each coulomb of charge passing through the battery.

10. The diagram shows the trace on an oscilloscope when an alternating voltage is applied to its input.

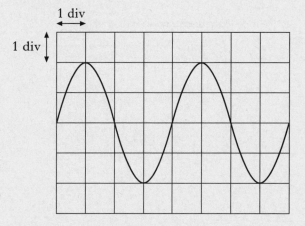

The timebase is set at $5 \, ms/div$ and the Y-gain is set at $10 \, V/div$.

Which row in the table gives the peak voltage and the frequency of the signal?

	Peak voltage/V	Frequency/Hz
A	7·1	20
B	14	50
C	20	20
D	20	50
E	40	50

[Turn over

11. A resistor is connected to an a.c. supply as shown.

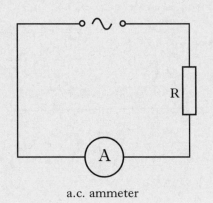

a.c. ammeter

The supply has a constant peak voltage, but its frequency can be varied.

The frequency is steadily increased from 50 Hz to 5000 Hz.

The reading on the a.c. ammeter

A   remains constant

B   decreases steadily

C   increases steadily

D   increases then decreases

E   decreases then increases.

12. An ideal op-amp is connected as shown.

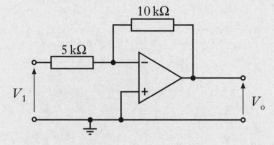

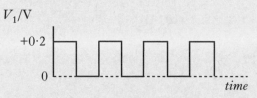

The graph shows how the input voltage, V_1, varies with time.

V_1/V

Which graph shows how the output voltage, V_o, varies with time?

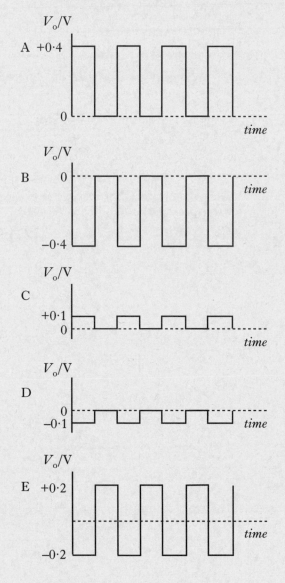

13. Which of the following proves that light is transmitted as waves?

    A   Light has a high velocity.

    B   Light can be reflected.

    C   Light irradiance reduces with distance.

    D   Light can be refracted.

    E   Light can produce interference patterns.

14. A source of microwaves of wavelength λ is placed behind two slits, R and S.

    A microwave detector records a maximum when it is placed at P.

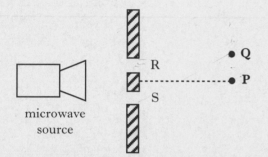

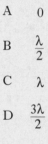

microwave source

The detector is moved and the **next** maximum is recorded at Q.

The path difference (SQ – RQ) is

    A   0

    B $\dfrac{\lambda}{2}$

    C   λ

    D $\dfrac{3\lambda}{2}$

    E   2λ.

15. A student makes five separate measurements of the diameter of a lens.

    These measurements are shown in the table.

Diameter of lens/mm	22·5	22·6	22·4	22·6	22·9

    The approximate random uncertainty in the mean value of the diameter is

    A   0·1 mm

    B   0·2 mm

    C   0·3 mm

    D   0·4 mm

    E   0·5 mm.

16. The value of the absolute refractive index of diamond is 2·42.

    The critical angle for diamond is

    A   0·413°

    B   24·4°

    C   42·0°

    D   65·6°

    E   90·0°.

[Turn over

17. Part of the energy level diagram for an atom is shown.

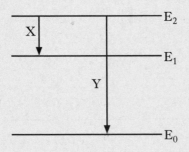

X and Y represent two possible electron transitions.
Which of the following statements is/are correct?

  I Transition Y produces photons of higher frequency than transition X.

  II Transition X produces photons of longer wavelength than transition Y.

  III When an electron is in the energy level E_0, the atom is ionised.

  A   I only

  B   I and II only

  C   I and III only

  D   II and III only

  E   I, II and III

18. The letters **X**, **Y** and **Z** represent three missing words from the following passage.

Materials can be divided into three broad categories according to their electrical resistance.

.............**X**............ *have a very high resistance.*

.............**Y**............ *have a high resistance in their pure form but when small amounts of certain impurities are added, the resistance decreases.*

.............**Z**............ *have a low resistance.*

Which row in the table shows the missing words?

	X	**Y**	**Z**
A	conductors	insulators	semi-conductors
B	semi-conductors	insulators	conductors
C	insulators	semi-conductors	conductors
D	conductors	semi-conductors	insulators
E	insulators	conductors	semi-conductors

19. Compared with a proton, an alpha particle has

  A   twice the mass and twice the charge

  B   twice the mass and the same charge

  C   four times the mass and twice the charge

  D   four times the mass and the same charge

  E   twice the mass and four times the charge.

20. For the nuclear decay shown, which row of the table gives the correct values of x, y and z?

$$^{214}_{x}\text{Pb} \longrightarrow {}^{y}_{83}\text{Bi} + {}^{0}_{z}\text{e}$$

	x	y	z
A	85	214	2
B	84	214	1
C	83	210	4
D	82	214	−1
E	82	210	−1

[Turn over for SECTION B on *Page ten*

SECTION B

Write your answers to questions 21 to 30 in the answer book.

Marks

21. To test the braking system of cars, a test track is set up as shown.

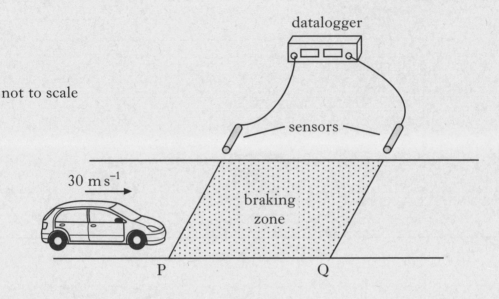

not to scale

The sensors are connected to a datalogger which records the speed of a car at both P and Q.

A car is driven at a constant speed of 30 m s^{-1} until it reaches the start of the braking zone at P. The brakes are then applied.

(a) In one test, the datalogger records the speed at P as 30 m s^{-1} and the speed at Q as 12 m s^{-1}. The car slows down at a constant rate of $9 \cdot 0 \text{ m s}^{-2}$ between P and Q.

Calculate the length of the braking zone.

2

(b) The test is repeated. The same car is used but now with passengers in the car. The speed at P is again recorded as 30 m s^{-1}.

The same braking force is applied to the car as in part (a).

How does the speed of the car at Q compare with its speed at Q in part (a)? Justify your answer.

2

Marks

21. (continued)

(c) The brake lights of the car consist of a number of very bright LEDs.

An LED from the brake lights is forward biased by connecting it to a 12 V car battery as shown.

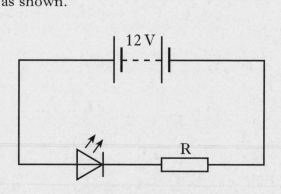

The battery has negligible internal resistance.

(i) Explain, in terms of charge carriers, how the LED emits light.   1

(ii) The LED is operating at its rated values of 5·0 V and 2·2 W.

Calculate the value of resistor R.   3

                   (8)

[Turn over

Marks

22. A crate of mass $40 \cdot 0 \, \text{kg}$ is pulled up a slope using a rope.

The slope is at an angle of $30°$ to the horizontal.

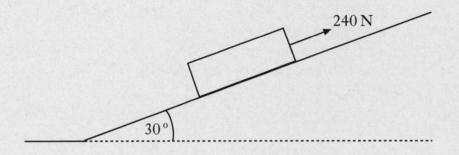

A force of 240 N is applied to the crate parallel to the slope.

The crate moves at a constant speed of $3 \cdot 0 \, \text{m s}^{-1}$.

(*a*) (i) Calculate the component of the weight of the crate acting parallel to the slope.   **2**

    (ii) Calculate the frictional force acting on the crate.   **2**

(*b*) As the crate is moving up the slope, the rope snaps.

The graph shows how the velocity of the crate changes from the moment the rope snaps.

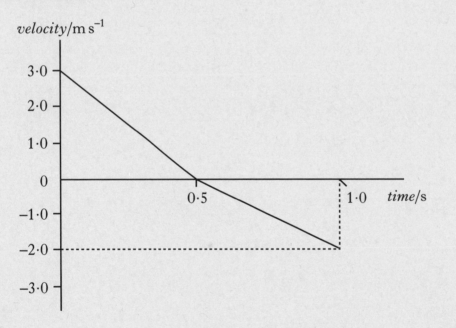

    (i) Describe the motion of the crate during the first $0 \cdot 5 \, \text{s}$ after the rope snaps.   **1**

Marks

22. (b) (continued)

(ii) Copy the axes shown below and sketch the graph to show the acceleration of the crate between 0 and 1·0 s.

Appropriate numerical values are also required on the acceleration axis.

2

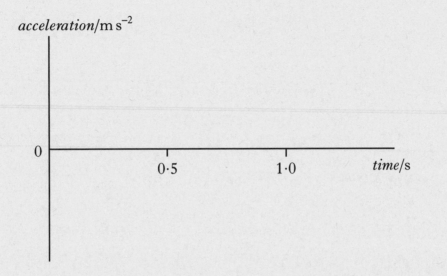

(iii) Explain, in terms of the forces acting on the crate, why the magnitude of the acceleration changes at 0·5 s.

2

(9)

[Turn over

Marks

23. A cylinder of compressed oxygen gas is in a laboratory.

(a) The oxygen inside the cylinder is at a pressure of $2 \cdot 82 \times 10^6 \, Pa$ and a temperature of $19 \cdot 0 \, °C$.

The cylinder is now moved to a storage room where the temperature is $5 \cdot 0 \, °C$.

  (i) Calculate the pressure of the oxygen inside the cylinder when its temperature is $5 \cdot 0 \, °C$.   **2**

  (ii) What effect, if any, does this decrease in temperature have on the density of the oxygen in the cylinder?

     Justify your answer.   **2**

(b) (i) The volume of oxygen inside the cylinder is $0 \cdot 030 \, m^3$.

    The density of the oxygen inside the cylinder is $37 \cdot 6 \, kg \, m^{-3}$.

    Calculate the mass of oxygen in the cylinder.   **2**

  (ii) The valve on the cylinder is opened slightly so that oxygen is gradually released.

    The temperature of the oxygen inside the cylinder remains constant.

    Explain, in terms of particles, why the pressure of the gas inside the cylinder decreases.   **1**

  (iii) After a period of time, the pressure of the oxygen inside the cylinder reaches a constant value of $1 \cdot 01 \times 10^5 \, Pa$. The valve remains open.

    Explain why the pressure does not decrease below this value.   **1**

                                                        (8)

Marks

24. Electrically heated gloves are used by skiers and climbers to provide extra warmth.

(a) Each glove has a heating element of resistance 3·6 Ω.

Two cells, each of e.m.f. 1·5 V and internal resistance 0·20 Ω, are used to operate the heating element.

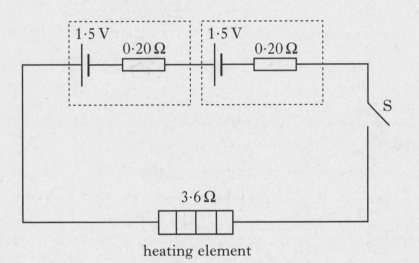

Switch S is closed.

   (i)   Determine the value of the total circuit resistance.   1

   (ii)   Calculate the current in the heating element.   2

   (iii)   Calculate the power output of the heating element.   2

(b) When in use, the internal resistance of each cell gradually increases.

What effect, if any, does this have on the power output of the heating element?

Justify your answer.   2

        (7)

[Turn over

Marks

25. (*a*) State what is meant by the term *capacitance*.　　1

(*b*) An uncharged capacitor, C, is connected in a circuit as shown.

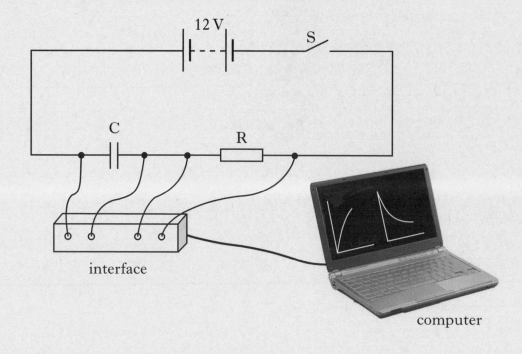

The 12 V battery has negligible internal resistance.

Switch S is closed and the capacitor begins to charge.

The interface measures the current in the circuit and the potential difference (p.d.) across the capacitor. These measurements are displayed as graphs on the computer.

Graph 1 shows the p.d. across the capacitor for the first 0·40 s of charging.

Graph 2 shows the current in the circuit for the first 0·40 s of charging.

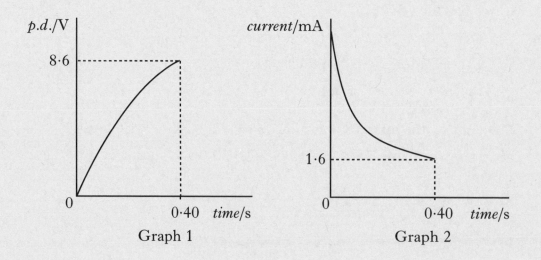

Marks

25. (b) (continued)

    (i)   Determine the p.d. **across resistor R** at 0.40 s.   **1**

    (ii) Calculate the resistance of R.   **2**

    (iii) The capacitor takes 2.2 seconds to charge fully.

          At that time it stores 10.8 mJ of energy.

          Calculate the capacitance of the capacitor.   **3**

(c) The capacitor is now discharged.
A second, identical resistor is connected in the circuit as shown.

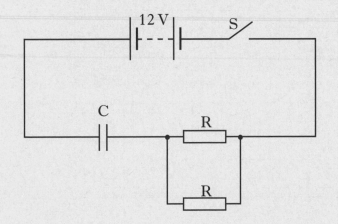

Switch S is closed.

Is the time taken for the capacitor to fully charge less than, equal to, or greater than the time taken to fully charge in part (b)?

Justify your answer.   **2**

       (9)

[Turn over

Marks

26. The graph shows how the resistance of an LDR changes with the irradiance of light incident on it.

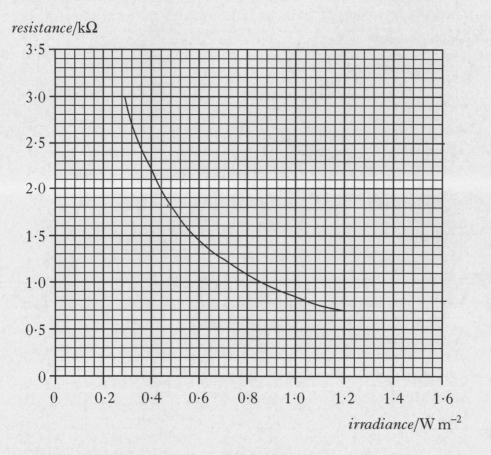

resistance/kΩ

irradiance/W m⁻²

(a) The LDR is connected in the following bridge circuit.

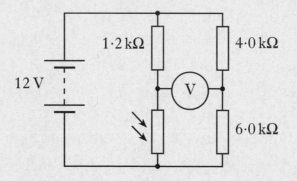

Determine the value of irradiance at which the bridge is balanced.

Show clearly how you arrive at your answer.

3

Marks

26.   (continued)

(b)   The LDR is now mounted on the outside of a car to monitor light level. It forms part of a circuit which provides an indication for the driver to switch on the headlamps.
The circuit is shown below.

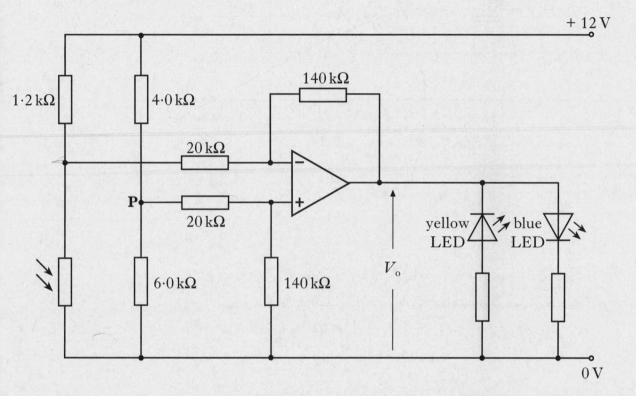

The LEDs inside the car indicate whether the headlamps should be on or off.

(i)   At a particular value of irradiance the resistance of the LDR is $2 \cdot 0 \, \text{k}\Omega$.

Show that the potential difference across the LDR in the circuit is $7 \cdot 5 \, \text{V}$.   1

(ii)   The potential at point P in the circuit is $7 \cdot 2 \, \text{V}$.

Calculate the output voltage, V_o, of the op-amp at this light level.   2

(iii)   Which LED(s) is/are lit at this value of output voltage?

Justify your answer.   2

(8)

[Turn over

Marks

27. (a) A ray of red light of frequency 4.80×10^{14} Hz is incident on a glass lens as shown.

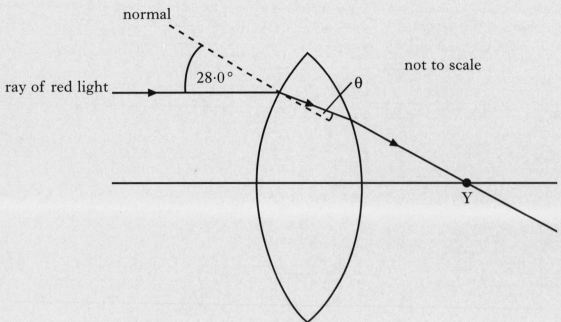

The ray passes through point Y after leaving the lens.

The refractive index of the glass is 1·61 for this red light.

  (i) Calculate the value of the angle θ shown in the diagram.   **2**

  (ii) Calculate the wavelength of this light inside the lens.   **3**

(b) The ray of red light is now replaced by a ray of blue light.

The ray is incident on the lens at the same point as in part (a).

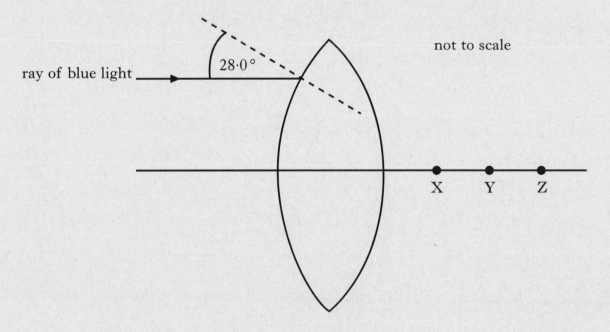

Through which point, X, Y or Z, will this ray pass after leaving the lens?

You must justify your answer.   **1**

    (6)

Marks

28. The diagram shows a light sensor connected to a voltmeter.

A small lamp is placed in front of the sensor.

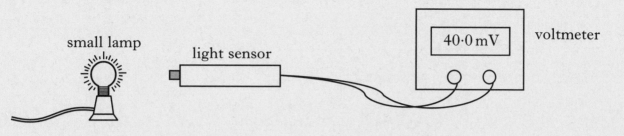

The reading on the voltmeter is 20 mV for each 1·0 mW of power incident on the sensor.

(a) The reading on the voltmeter is 40·0 mV.

The area of the light sensor is $8·0 \times 10^{-5} \, m^2$.

Calculate the irradiance of light on the sensor. **3**

(b) The small lamp is replaced by a different source of light.

Using this new source, a student investigates how irradiance varies with distance.

The results are shown.

Distance/m	0·5	0·7	0·9
Irradiance/W m^{-2}	1·1	0·8	0·6

Can this new source be considered to be a point source of light?

Use **all** the data to justify your answer. **2**

(5)

[Turn over

Marks

29. To explain the photoelectric effect, light can be considered as consisting of tiny bundles of energy. These bundles of energy are called photons.

   (a) Sketch a graph to show the relationship between photon energy and frequency.   1

   (b) Photons of frequency $6 \cdot 1 \times 10^{14}$ Hz are incident on the surface of a metal.

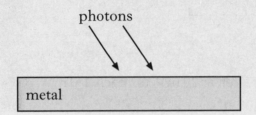

   This releases photoelectrons from the surface of the metal.

   The maximum kinetic energy of any of these photoelectrons is $6 \cdot 0 \times 10^{-20}$ J.

   Calculate the work function of the metal.   3

   (c) The irradiance due to these photons on the surface of the metal is now reduced.

   Explain why the maximum kinetic energy of each photoelectron is unchanged.   1

   (5)

Marks

30. (*a*) A technician is carrying out an experiment on the absorption of gamma radiation.

The radioactive source used has a long half-life and emits only gamma radiation. The activity of the source is 12 kBq.

    (i) State what is meant by an *activity of 12 kBq*.   **1**

    (ii) The table shows the half-value thicknesses of aluminium and lead for gamma radiation.

Material	Half-value thickness/mm
aluminium (Al)	60
lead (Pb)	15

The technician sets up the following apparatus.

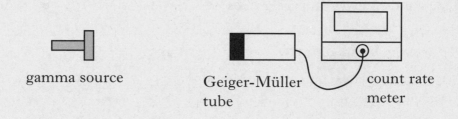

gamma source   Geiger-Müller tube   count rate meter

The count rate, when corrected for background radiation, is 800 counts per second.

Samples of aluminium and lead are now placed between the source and detector as shown.

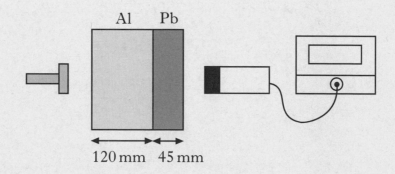

Al   Pb

120 mm   45 mm

Determine the new corrected count rate.   **2**

[Turn over for Question 30 (*b*) on *Page twenty-four*

Marks

30. (continued)

(b) X-ray scanners are used as part of airport security. A beam of X-rays scans the luggage as it passes through the scanner.

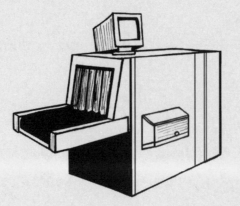

A baggage handler sometimes puts a hand inside the scanner to clear blockages.

The hand receives an average absorbed dose of $0.030\,\mu Gy$ each time this occurs.

The radiation weighting factor for X-rays is 1.

  (i) State the average equivalent dose received by the hand on each occasion.   **1**

  (ii) The occupational exposure limit for a hand is $60\,\mu Sv$ per hour.

    Calculate how many times the baggage handler would have to put a hand into the scanner in one hour to reach this limit.   **1**

        (5)

[END OF QUESTION PAPER]

[BLANK PAGE]

[BLANK PAGE]

[BLANK PAGE]

[BLANK PAGE]